The Greening of Georgia

The Improvement of the Environment in the Twentieth Century

by R. Harold Brown

ISBN 0-86554-789-0
MUP/H596

© 2002 Mercer University Press
6316 Peake Road
Macon, Georgia 41210-3960
All rights reserved

First Edition.

Book design by Mary-Frances Burt
Burt & Burt Studio

The paper used in this publication meets the minimum requirements of American National Standard for Information Sciences—Permanence of Paper for Printed Library Materials, ANSI Z39.48-1984.

Library of Congress Cataloging-in-Publication Data

Brown, R. Harold.
The greening of Georgia : the improvement of the environment in the twentieth century
R. Harold Brown — 1st ed.
p. cm.
Includes bibliographical references.
Contents: 1.The need for good news—Changes in land use—What was the landscape when the settlers came?—Clearing of the land—Changing agriculture—Diversification—Reforestation—Wetlands—Urbanization—Restoring the land—Erosion: the great disaster—The uplands were wasted—The streams were spoiled—Conserving the soil—Governmental and civic action—What are the results?—Covering the bare places— Reducing the silting of streams and lakes—How did it happen?—Cleaning of the water—Water quality in the past—Clearing the streams—Reducing sewage pollution—Reducing nitrogen and phosphorus—Removal of industrial pollutants and pesticides—Overall improvements in water quality—Protection of stream quality— Clearing the air—Smoke and smog—How clean was the air back then?—Monitoring the air—Dust, dirt, and particles—Sulfur dioxide and acid rain—Ozone—Lead poisoning—Carbon monoxide—Miscellaneous pollutants—Visibility—Air pollution and health—Restocking the wildlife—Wildlife in the past—Game species—Deer—Wild turkey—Quail and dove—Rabbits and squirrels—Furbearers—Alligator—Waterfowl—Nongame species—Songbirds—Fire in the forest—Fish—Seafood—Realities and perceptions—Summing up.

I. Georgia—Environmental conditions.
2. Environmental protection—Georgia.
I. Title.

GE155.G4 B76 2002

333.7'2'09758—dc21   2002002616

TABLE OF CONTENTS

	Acknowledgements	v
Chapter 1.	**The Need for Good News**	1
Chapter 2.	**Changes in Land Use**	15
	What Was the Landscape When the Settlers Came?	15
	Clearing of the Land	18
	Changing Agriculture	20
	Diversification	26
	Reforestation	31
	Wetlands	41
	Urbanization	49
Chapter 3.	**Restoring the Land**	59
	Erosion: The Great Disaster	59
	The Uplands Were Wasted	61
	The Streams Were Spoiled	70
	Conserving the Soil	77
	Governmental and Civic Action	79
	What Are the Results?	84
	Covering the Bare Places	85
	Reducing the Silting of Streams and Lakes	92
	How Did It Happen?	114
Chapter 4.	**Cleaning of the Water**	119
	Water Quality in the Past	119
	Clearing the Streams	124
	Reducing Sewage Pollution	128
	Reducing Nitrogen and Phosphorus	138
	Removal of Industrial Pollutants and Pesticides	157
	Overall Improvements in Water Quality	176
	Protection of Stream Quality	182

Chapter 5.	**Clearing the Air**	**189**
	Smoke and Smog	*189*
	How Clean Was the Air Back Then?	*193*
	Monitoring the Air	*199*
	Dust, Dirt, and Particles	*202*
	Sulfur Dioxide and Acid Rain	*209*
	Ozone	*223*
	Lead Poisoning	*236*
	Carbon Monoxide	*239*
	Miscellaneous Pollutants	*242*
	Visibility	*245*
	Air Pollution and Heath	*249*
Chapter 6.	**Restocking the Wildlife**	**255**
	Wildlife in the Past	*255*
	Game Species	*259*
	Deer	*259*
	Wild Turkey	*265*
	Quail and Dove	*268*
	Rabbits and Squirrels	*276*
	Furbearers	*278*
	Alligator	*283*
	Waterfowl	*284*
	Nongame Species	*287*
	Songbirds	*290*
	Fire in the Forest	*303*
	Fish	*306*
	Seafood	*315*
	Realities and Perceptions	*321*
Chapter 7.	**Summing Up**	**333**
	References	339

ACKNOWLEDGEMENTS

The writing of this book would have been more difficult and much less complete without the help of many people. There are too many to mention by name. I am indebted to the staff at the University of Georgia Libraries who helped acquire materials from several sources. Personnel in the Georgia Department of Natural Resources, the US Environmental Protection Agency, the US Geological Survey, and other governmental agencies have supplied reports and information. Several other individuals supplied data and information, and some have been acknowledged in places where the information is used. Several friends and colleagues have read portions of the manuscript. I am grateful for the help of all who supplied materials and suggestions and the book has been greatly improved because of them. While I have eagerly accepted the help and suggestions of others, any errors of judgement and fact in the book are mine. I hope they are few.

I am especially indebted to my wife, Pat, for her patience and understanding during the six years of long hours studying in my office and in the library and sitting before my computer, when I was supposed to have been "retired."

Eternal Chattahoochee

Dear little baby mountain stream, trickling down the valley, clean,
And growing bold by Habersham, and over boulders tumbling down,
Promenade by farm and town.
    Proud Chattahoochee, Georgia's crown.
When logs were floated on your breast, from river's edge and mountain crest;
When heavy boats plowed slow and hard, past smoking stack and lumberyard,
You gave your gift without regard,
    Receiving, asking, no reward.
Then cotton billowed in the fields, and rain-swept earth flowed down the hills.
You bore that wretched, rusty load from untold gully, ridge, and knoll,
And land and river paid the toll.
    Soiled and sullied, wounded soul.
On your broad banks new cities gleam, the profiles of a modern dream.
They notice not the shades of gray, the dust and dirt of every day,
That wind and river wash away.
    Redeeming, cleansing, Georgia clay.
When man first saw your waters flow through gentle hills and mild plateau,
And watched amazed from wooded glen, your rugged crest fade round the bend,
Your fate was changed forever, then.
    Native treasure, modern friend.
From ancient mountain cove you came, born of prehistoric rain,
In foggy, early morning mist, by noonday sun and shadow kissed.
Through ages to that great abyss;
    Fleeting sorrow, endless bliss.
We've heard from those who say you'll die, but it's a weak and easy lie.
You'll run your course when they are gone, with countless others you have known,
Who walked your banks and waved you on.
    Dear everlasting river song.

—Harold Brown

CHAPTER ONE

The Need for Good News

Why tell the good news about the environment of Georgia? Why use such an optimistic title? Is there really a positive side to the environmental story frequently told in newspapers and books and on television? Good news about the environment is seldom heard, but bad news is standard fare, even as problems have become smaller. In contrast, news about advances in some areas that affect our health and happiness, for example, medical science, appears regularly. Cures are hailed even when they are only successful on laboratory animals and long before they are tried on humans. A recent newspaper story about the teaching of environmental issues in public schools raised the question, what should school children be taught? It is not just the media that skews coverage toward environmental failure. A national television program recently asserted that school teachers were teaching only about environmental problems and none of the progress.[1] It seems fairly obvious that school children should learn about the improvements as well as the problems. It is more important that they learn enough science to make better-informed judgments later about this scientific subject. Perhaps that is too much to hope for in an atmosphere heavily biased toward advocacy yet very limited in analysis.

Historically, the environmental movement is steeped in a tradition of bad news. It appears that there is little room for good news, perhaps because of fear that it dilutes the efforts to clean the environment. In fact, a friend of mine told me that writing a book of good news might be detrimental to the effort to clean up Georgia's environment. However, apathy does not arise from good news, but rather the lack of it. This imbalance in environmental coverage is not due to a lack of progress in the environment. The progress is evident even in the health of the American people. The average life span has increased dramatically in this century. Life expectancy in the United States rose from forty-seven years for a person were born in 1900 to sixty-eight years if born in 1950 and seventy-seven years if born in 2000. As late as 1940, only 5 percent of Georgia's population were over sixty-five years old. In 1998, 10 percent were over sixty-five, not because we are a retirement state, but because we are living longer.

Increases in the American life span are not just because of the improvements in medical diagnosis and treatment, but also because our environment is less hazardous. These changes are obvious in the comparisons of conditions early and late in the twentieth century. In 1940, 59 percent of Georgia dwellings and 90 percent of rural farm dwellings had no indoor water system. There were no flush toilets in 75 percent of Georgia homes. Cleanliness has improved. Common ailments among children such as skin boils, eye infections (sties or "sore eyes"), and hookworms (intestinal parasites resulting mainly from barefoot contact with contaminated soil) are no longer common. More serious diseases were also related to the environment. Typhoid fever, dysentery, and malaria were common in Georgia early in the century. Ingestion of food or water contaminated with the bacterium *Salmonella typhosa* caused typhoid fever, and malaria was transmitted by mosquitoes associated with swampy areas we now cherish as wetlands. Contaminated food and water frequently caused dysentery.

Environmental improvement is obvious in the appearance of the countryside. The landscape in the early part of theis twentieth century was decidedly more stark and bare than now. Roads were unpaved, and bare soil was one of the dominant features of the countryside. In rural Georgia, as late as the 1940s, home yards (not lawns) were swept bare rather than having mowed grass as they do now. On the wall in my office is an aerial photograph of Athens, Georgia and surroundings taken in 1934, and perhaps the most striking feature of the picture is the lack of forest. The land was cultivated right to the edge of the Oconee River. Cultivation of the rolling Piedmont, and to a lesser extent the Coastal Plain of Georgia, exposed it to extensive erosion of its soil into the streams, as we shall see later.

Compare the roadsides of today with those of anytime before 1950 and the contrast is remarkable. Rutted roadbeds of the early 1900s were rough and dusty in dry weather and quagmires in rainy periods. In wet weather they were treacherous for the driving of automobiles, and washed-out bridges and overflow of low-lying roads prevented many a trip to school or town. C. C. Thomas, president of the Georgia Highway Association in 1926, said "And yet, Georgia, the Empire State of the South, with her vast resources and opportunities undeveloped, is still in the mud and her main highways impassable during many months of the winter season." Bare red backslopes of the north Georgia roads were not as pleasing as the grassy or tree-lined slopes of today. Conditions of streets in the cities and towns were not much better. As late as the 1940s, I recall seeing cattle and pigs roaming the streets of small towns in Middle Georgia.

Even though today's news is about problems and accidents, it is obvious to many people that the environment has improved. In a telephone survey of 250 residents in each of four counties in the late 1980s, the percentage of persons who thought the environment had improved over the last ten years ranged from 15 percent in Glynn to about 40 percent in Oglethorpe and Emanuel Counties.[2] When responses from all four counties were averaged, the opinions were about evenly divided among "Improved," "Worsened," and "Same." This survey may not be representative of Georgians, because it was small and polled people in mostly rural counties, but it shows that in spite of the bad news, about one-third of the people in these counties believe the environment has gotten better. The percentage of people in metropolitan areas who think the environment has gotten better may be smaller for two reasons. First, the publicity about environmental problems is more intense in the metropolitan press, and second, environmental problems are worse in metropolitan areas. However, it is paradoxical that in areas where negative publicity is most persistent, the improvements have been most dramatic. The cleaning of water and air in Atlanta and Georgia's smaller cities is much more dramatic than in rural areas where it was never as polluted.

Whether the state of the environment is good or bad, depends on one's perspective and the basis for comparison. What are our environmental goals? Do homemakers, farmers, office workers, teachers, and construction workers perceive the environment as equally bad? Is the environment better or worse than it used to be? Is it as good as it should be? Should we strive to restore the environment to that of pre-colonial times, or even to what it was at the beginning of this century? What was it like at those times? What is it like today? A rational approach to remedies for environmental problems should take account of such questions.

News or popular press accounts of the environment are partial descriptions of its problems. Such informative accounts are not very productive if there is no basis for the descriptions of its improvement or degradation. That is to say, the state of the environment is relative. If it is said to be poor, there should be reference to some time when it was ideal, or at least better. There were times when the environment was better than now, if one's reference and preference is the imaginary pristine condition before man influenced it. Of course, humans cannot live in an environment that they do not influence. Although there is now advocacy for reduced human influence on the environment, there has been in the past emphasis on taming nature.

The focus of most discussion in the media and even in current scientific literature is on the "poor state" of the environment. Occasionally, there is good news about the environment, but in many cases it is incidental to a larger complaint. In a front-page story in April 2000 entitled "Earth Day: Progress 30 years later?" there were recitations of cleaning the Chattahoochee, recovering the bald eagle, and the increased efficiency and reduced pollution of today's cars.[3] This accounted for about 10 percent of the article and was a prelude to the following statements. "…Georgia is still being sullied, eroded and paved over at an alarming pace: Metro Atlanta is losing about 50 acres of tree cover per day for new subdivisions…" "The region's dirty air seems intractable." "…more than 10 percent of plant and animal species in the state are imperiled to one degree or another." The article went on to describe the Chattahoochee River: "…the waterway perennially ranked in the top 10 as one of America's most endangered rivers…." The bad news in this article outweighs the good ten to one.

The problems of contamination of air, water, and food, and conservation of natural resources are a national concern and are frequently highlighted, but analysis in the press is usually inadequate or absent. Clean water is a personal and social issue, but is also a political one. Communities downstream are naturally suspicious of the treatment of water by upstream neighbors. In a lead editorial in the *Atlanta Constitution* of 4 February 1996, about pollution of the Chattahoochee, the headline was "Hooch Has Had Enough." The editorial was about the delay by the city of Atlanta in removing phosphorous and other elements from water released into the Chattahoochee. An accompanying chart showed phosphorus in the water discharged from treatment plants had decreased from about 5.5 parts per million in 1988 to less than 1 part per million in 1995. The editorial did not mention the chart or the 80 percent decrease in phosphorus over eight or nine years, nor did it say how close Atlanta was to cleaning the water before it was released. The tone of the article was decidedly negative and it gave little information about the problem.

Hyperbole is all too common in discussions of the environment. Vice President Al Gore wrote in the introduction of a 1994 reissue of Rachel Carson's book *Silent Spring* that "Despite the power of Carson's argument, despite actions like the banning of DDT in the United States, the environmental crisis has grown worse, not better.... Since the publication of *Silent Spring*, pesticide use on farms alone has doubled to 1.1 billion tons a year, and production of these dangerous chemicals has increased by 400 percent." It is certain that the environment is not in "crisis" and the great bulk of the pesticides used on US farms are not "dangerous." In contrast to the vice president's statement, in his 1999 state of the union address, President Clinton said, "Our environment is the cleanest in 25 years." The environment could hardly have recovered so dramatically in four or five years, but at least for Georgia and doubtless for the nation, President Clinton had it right.

In today's critique of the environment, three main areas are usually emphasized. First, global warming is predicted as a result of the increase of carbon dioxide and other gases in the atmosphere. Second, there is concern about contamination of water, air, and food by pesticides and other by-products of modern living. Third, recycling rather than disposal of certain items is emphasized to save the raw materials, energy, and disposal space. Global warming is the most controversial and least certain of these. The other two areas of concern should be much less in question because of three or four decades of emphasis and progress.

One of the problems in telling the good news about Georgia's environment is that it must be compared to the past. This is a problem because for many the past is passé, of little relevance to the present. It is a hard sell to young people, or seniors for that matter, that poor conditions before they were born should make them feel better about present problems. The past is relevant, however, because therein lies the root of the problems (or solutions) of the present. As said so well by Reeves and McCabe, "In biology, as in other fields, history is the genesis of the present and precursor of the future. It serves as the canon against which prevailing events, circumstances, standards, and thoughts are measured and judged. History is perspective."[4] The definition of "improvement" (or "degradation") requires that we compare today's environment against that of yesterday.

Frequently, progress, or lack of it, is judged by the relative standing of states. If Georgia is found to stand twenty-fifth among the fifty states in education or some other measure of progress, that is taken as justification for increased funding or some other means to improve the standing. Georgia must always be competitive and strive to improve the lot of its citizens, but relative standing is not necessarily a measure of progress. Most students and parents do not view graduation in the middle of the class at the University of Georgia or Georgia Tech as something less than success; in fact it is good news. While I do not know the relative standing of Georgia's environment among the United States, I would conclude by analogy that we have progressed from early elementary through high school and perhaps a couple of years of college. The decision now is how do we make it to graduation, and should we pursue advanced degrees.

The goal of this book is very limited and straightforward. It does not attempt to give an exhaustive description of the state of the environment, to argue the responsibility for its past or present state, or to espouse any course of action. The goal is to describe the improvements in Georgia during the twentieth century with enough documentation to try and convince the reader that the trends are not only real, but also substantial. Detrimental trends will not be ignored, but the emphasis will be on the good news, because it so far outweighs the bad, and the bad news has had more than its share of press. Because perception of the state of the environment is based on one's knowledge, experience, and outlook, great diversity in our perception is inevitable. However, a balanced view is desirable and the improvements made during the twentieth century need to be described.

The trends of knowledge, awareness, and technology in the twentieth century are such that the main part of the story begins in the middle of the century and the environmental improvement occurs mainly in the last half or last quarter of the century, depending on the specific problem. During the first half of the century, Georgia was largely an agricultural society and the most serious problems tended to be related to agriculture. After World War II increasing movement of the population to the cities and increasing industrialization created a different set of environmental problems, and the problems moved to the cities. While increasing population and industrialization in the cities increased pollution problems, increased education, affluence, and technology allowed for solutions of those problems.

There are at least three benchmarks for the current relative health of the environment; the condition before humans influenced it, the condition at some arbitrarily fixed time in the past, and the condition that is ideal for the society in which one lives. All of the benchmarks are subjective and debatable. The first is impossible to match because humans influence any environment in which they live. Many feel that the environment in Georgia and the rest of the United States before colonization by Europeans was close to that before humans affected it. However, there is little knowledge of how North American Indians influenced the environment. By some accounts they caused large changes. Indians regularly burned the forests and grasslands in some areas. There is also some evidence that Indians cultivated rather larger areas than previously thought, mostly along the larger streams. Therefore, one question is how green did Georgia used to be? For most of the past we can only judge from inadequate records.

An increase in the number and cultural activity of European settlers certainly caused a major change in the landscape. The clearing of the forests, cultivation of crops on a much larger scale than the Indians had done, and the building of roads, were major factors in changing the landscape. Industrialization of Georgia with the large increase of pollutants released to the air and water decreased the quality of both. The increased use of agricultural and household pesticides and the lack of knowledge about their harmful effects caused further damage to the environment. A small example of the lack of knowledge about the danger of pesticides was the use of a DDT solution to soak the lids of wild duck nesting boxes on the Piedmont National Wildlife Reserve.[5] This was done to reduce the hazard of wasp stings for workers checking nesting progress. Another example is the description by Jimmy Carter of insect control by mopping of cotton plants with a mixture of arsenic and molasses as a young boy.[6] He recalls wearing the chemical-soaked clothes day after day before they were washed. These are not isolated examples and they show that the problems posed by advanced technology for environmental quality and health were largely unrecognized before the 1960s and 1970s.

The need for good news about the environment can be supported in several ways, but four are most obvious. First, if truth is the scale on which we balance the information given to the public, a large dose of good news about the environment is justified and truth lends credibility. It is dubious policy for government or other organizations to claim no progress, and worse, a deterioration, after expenditure of vast sums on the environment since the 1960s. In fact, annual reports of government agencies regularly contain the good news. The problem is that the news seldom spreads beyond government reports. Interest groups make conflicting claims about specific environmental problems usually with little or no objective numbers, and comprehensive reports on environmental progress are almost never publicized. Credibility is a valuable asset for government agencies and for others charged with responsibility for protecting the environment. Openness, accomplishment, and straight talk strengthen it. The populace will not forever believe a worsening of the environment if the claims conflict with easily perceived improvements such as more plentiful fish and wildlife, cleaner water, more beautiful roadsides, and fresher air. The Environmental Protection Agency listed as the seventh of ten goals in its 2000 budget summary the "Expansion of Americans' Right to Know About Their Environment."[7] The report explained, "The purpose of this goal is to empower the American public with information, enabling them to make informed decisions regarding environmental issues in their communities." Although "right-to-know" laws and regulations have dealt mostly with the need to be informed about the danger of environmental pollution, there is an equal "right-to-know" about the progress that has been made. They are two sides of the same coin. Knowledge about progress needs some expansion. A recent survey found that "90 percent of Americans consider environmental factors such as pollution, waste and chemicals important contributors to disease."[8] If so many Americans see the environment as a threat to something so important as health, good news about the environment must be told.

But a second and perhaps equal obligation for the telling of the good news is that there must be some public accounting of the large sums of money spent to solve environmental problems. Figure 1.1 shows that the annual budget of the Georgia Environmental Protection Division increased by more than 2.5 times, after adjusting for inflation, from the early 1970s to the 1990s.[9] The EPD budget is only a small portion of the increased cost of cleaning or maintaining the environment; the federal government, other state agencies, and private industries have also spent large and increasing sums. The Council on Environmental Quality reported that the national cost of pollution abatement and control rose from about $17 billion in 1972 to $122 billion in 1994 (Figure 1.1).[10] Private industry paid the greatest share; 62 percent in 1993.[11] It was estimated in 1988 that $750 million had been spent in Atlanta alone to reduce a small component of total pollution, the release of volatile organic compounds, which contribute to ozone buildup in the air.[12] Of course, ultimately citizens pay the cost. After the spending of hundreds of billions of dollars to solve environmental problems what are the results? Justification of the spending of so much money requires good news.

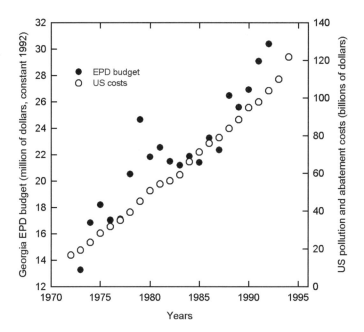

Figure 1.1. Increases in the budget of the Georgia Environmental Protection Division (Kundell and Dorfman, 1992) and US costs of pollution control and abatement (Council on Environmental Quality, 1997) from the early 1970s to the 1990s. EPD funding has been adjusted for inflation to the 1992 value of the dollar.

The third reason for good news is related to the second. From the beginning, a main justification for spending money to clean up the environment has been economic. A view frequently expressed is that environmental clean-up does not cost—it pays. It pays in better health of citizens, lower medical bills, and fewer days of lost work. It pays in greater productivity of our farms, forests, and fisheries. The national benefits of reducing air pollution alone in the ten years following the Clean Air Act of 1970 was estimated at $21.4 billion.[13] If the tremendous investment in cleaning and maintaining the environment in the latter half of the twentieth century has turned a profit in better health and increased productivity, then it is certainly important to discuss the returns from our environmental investments.

A fourth justification for telling the good news is to reinforce the efforts to further clean the water and air. Modern psychology has stressed positive reinforcement as a powerful way of improving behavior. The environmental movement accomplished nothing more important than the education of the public about the dangers of environmental carelessness. But just as no organization motivates its members by constant bad news, no movement can persist if fueled only by failure, or perception of failure. Substantial progress toward the goal of an improved environment is necessary to sustain enthusiasm for the fight. Ask any football coach whether motivation is easier with success or in its absence. If we believe positive reinforcement is a way of encouraging, not just compliance, but enthusiasm, then a "pat on the back" for environmental regulators who have enforced the rules, for companies who have reduced pollutants, and for citizens who have worked hard at recycling and car-pooling should help the environmental movement. A notable group of veiled heroes could be recognized if the good news were told. The mammoth change brought on by the environmental successes in Georgia reflects a great deal of credit on many people. But our civic debt and their conscience and hard work go largely unheralded because bad news obscures any accomplishment.

If the news media and environmentalists continue a campaign of nagging negativism about the environment in the face of positive results, how will Georgians get the good news? Who will applaud the industries that have reduced pollution to the lowest level in generations? Who will salute the scientists and technicians who have developed the knowledge and technology to solve pollution problems without jeopardizing economic progress? Who will say to the idealistic young people of the last forty years that their protests against pollution were not empty? This description of the improvement of Georgia's environment is presented as a tribute to all of them.

[1] Stossel, 2001.
[2] Odum and Turner, 1987.
[3] Seabrook, 2000a.
[4] Reeves and McCabe, 1993.
[5] Almand, 1965.
[6] Carter, 2001.
[7] US Environmental Protection Agency, 1999a.
[8] Associated Press, 2000.
[9] Kundell and Dorfman, 1994.
[10] Council on Environmental Quality, 1997.
[11] Council on Environmental Quality, 1994.
[12] Georgia Department of Natural Resources, 1988.
[13] Seneca, 1987.

CHAPTER TWO

Changes in Land Use

"Hitler's taunt that no democracy uses its land decently, while true of our past must be proven untrue in the years to come."

—*Aldo Leopold in a seminar during World War II*[1]

What Was the Landscape When the Settlers Came?

Many accounts of the settlement of the eastern seaboard of North America speak of continuous forests cleared by settlers, and it is widely assumed that the continuous forests existed for millennia. Peterson (1979) referred to "an unbroken wilderness of ancient trees." Although the land was mostly covered with forests, the extent of cleared land has probably been underestimated. Early descriptions of the Georgia landscape refer to great open spaces. Rostlund (1957) relates two references to open land from quite different periods. First he states that the Spanish explorer, Garcilaso, in the sixteenth century crossed the Flint River in southwestern Georgia and came to "clear land [tierra limpia] with many cornfields." Farther north but still west of the Flint, probably in Lee County, "they doubled their marches and could easily do so, for the land was level without woods, mountains, or rivers." Garcilaso also says that the land along the Ocmulgee River was "fertile and abundant, with fine forests and clearings [rasos]." Rostlund (1957) also quotes from William Bartram in June 1776, that they camped for the night at the Oconee River in "a delightful grove of oak, ash, mulberry, hickory, black walnut, elm, sassafras, locust, etc. This grove extended into an extensive, green, open, level plain, consisting of old Indian fields and plantations, stretching to a very great distance."

The slash and burn agriculture of the southeastern Indians may have resulted in large areas of cleared land in what was to become Georgia and the amount of land cleared probably changed over the centuries. Indians cleared land to grow crops and were required to continually clear new lands because of the rapid decrease in fertility of cleared soils. If the lands had been left alone, they would have quickly reverted back to forest. But according to Rostlund (1957),

> the land was not left undisturbed. The Indians customarily burned over not only the woodlands but the open tracts as well, which also became favorite hunting grounds; and this burning, to judge from the old reports, was so common and widespread that it is highly improbable that any large part of the cleared and abandoned land had a chance of reverting to forest. On the contrary, it is far more likely that the area of this type of open country was steadily increasing, and since this aboriginal deforestation had been in progress for a long time, for millenia rather than centuries, the upshot is that the open country made by men must have constituted a very considerable part of the old Southeast.

Some authors believed that the burning of the forest by Indians set the stage for invasion by buffalo that are known to have been numerous in the southeast. Although forest with small clearings is favored habitat for the white-tailed deer, open grazing land is required for buffalo. The amount of clear or open grazing land at the time of settlement must have been considerable judging from the extent of buffalo in the area. Rostlund (1960) gives several observations of early explorers in what was to be Georgia that indicates extensive herds. He quotes one of Oglethorpe's Rangers on 31 July 1716, west of the Oconee River, probably in Laurens County "We killed two buffaloes, of which there are abundance, we seeing [sic] several herds of sixty or upwards in a herd." Thomas Spaulding is quoted as relating this about the area of McIntosh County in the 1740s: "Colonel William McIntosh...has often told me that he has seen ten thousand buffaloes in a herd between Darien and Sapelo River." According to Rostland (1960), buffalo began migrating into southeastern North America in the mid-1500s and the population reached a peak about 1700. Rostlund also concludes that "burning the woods, practiced no doubt for many centuries before 1500 AD, favored the spread and increase of the grasses, and we can therefore say that the buffalo pastures of the Southeast were in large measure made by man." So the presence of large numbers of buffalo was evidence of widespread clearing and burning in earlier centuries.

Such revelations do not negate the notion that the aboriginal land of Georgia was mostly forest, but they cast doubt on the continuous nature of the forests. Clearings made and maintained by the Indians served to attract certain game species, made hunting of game easier, increased visibility for protection from enemy tribes, and were also used for cultivation of crops. Prunty (1965) postulated that Indians and the early settlers fought over clearings in the forest. Although the amount of cleared land when the white man came was probably greater than has been widely assumed in modern times, it was also probably less than in earlier centuries. Archaeological evidence indicates that Indian populations in eastern North America had decreased drastically in the century or two before Columbus. So the forests had time to partially recover from clearing by Indians. These considerations led Rostlund (1957) to conclude that "Paradoxical as it may seem, there was undoubtedly much more 'forest primeval' in 1850 than in 1650."

Clearing of the Land

Whatever the extent of clearing and cropping the Indians may have practiced, it was negligible compared to the clearing that occurred after settlement by the Europeans. General Oglethorpe established Georgia's first colony on the coast in 1733, but there was little expansion away from the coast until much later. Late in the century there was large-scale movement into the Piedmont of Georgia from the Carolinas, Virginia, and states further north. According to Coulter (1965), there was a large influx of settlers from Virginia into the Broad River Valley above Augusta in the late 1700s with the peak period being from 1783 to 1790 and by the early 1800s the land was nearly all settled. The census of 1810 showed 45,000 people in the Broad River Valley. The town of Savannah, by comparison, had only 5,215.[2] But it was not just the Broad River Valley that filled up; the whole of the area bordering South Carolina was settled.

Settlement proceeded rapidly westward across the Piedmont. In 1785, Elijah Clarke and John Twiggs signed a treaty in which the reluctant Creek Indians agreed to confirmation of a previous cession of lands lying between the Ogeechee and Oconee Rivers.[3] In the treaty the Creeks also ceded the territory lying east of a line running from the fork of the Oconee and Ocmulgee Rivers to the head of the St. Mary's River on the Florida border.[4] A map of the Piedmont presented by Trimble (1974) shows a well-populated region east of the Oconee River in 1810 but almost no settlement west of the Oconee. In 1805, however, the Indians ceded the land between the Oconee and Ocmulgee Rivers to the state of Georgia, and by the 1830s the Piedmont in Georgia was completely settled. At mid-century, the population was as great or greater in the western Piedmont than in the east.[5]

The land settled by the newcomers was soon cleared of forest and cropped with tobacco, cotton, corn, and a variety of minor crops. In the early clearing of the Piedmont, some of the trees were used in building and for fuel, but most were cut and burned to make room for cultivation. So following the tide of population, there was a wave of clearing of land that proceeded across the Georgia Piedmont from east to west. Of course, not all of the trees were cut in the beginning, but after clearing there was a rapid decline in soil fertility and new fields were needed for cultivation. Statistics on the extent of clearing during settlement are not available, but some early observations reveal the devastation that resulted.[6]

The vast Coastal Plain region was put into cultivation much later than the Piedmont. In the westward rush for land, that area south of the Fall Line from the coast to the Chattahoochee River was considered "sterile pine barrens" more suited for livestock grazing and hunting than for commercial agriculture.[7] However, by the 1860s much of the western part of the plain around the Flint River was settled and had larger and more prosperous farms producing more cotton than the old Cotton Belt. The coastal strip had grown in population and in the 1850s Savannah was still the largest city in the state with about 22,000 population. The interior of the Coastal Plain was the last to be put under cultivation and its settlement waited until the late 1800s for the coming of railroads and the widespread cutting of timber by large companies that depended on the railroads.

Because of a combination of poor soil, land scandals, and speculation that allowed large parcels to be tied up, that portion of the upper Coastal Plain wiregrass region between the lower reaches of the Oconee and Ocmulgee Rivers was the last to be populated.[8] Whereas the more northern lands between the Oconee and Ocmulgee Rivers were settled by cotton planters, the lower four counties, Laurens, Pulaski, Montgomery, and Telfair were settled mainly by self-sufficient farmers and livestock herders. In 1860, almost two generations after they were opened for settlement, the four counties had fewer than six inhabitants per square mile and just after the Civil War only about 8 percent of the land was cultivated.[9]

Late in the 1800s there was a mass movement of people into the interior of the Coastal Plain spurred on by the building of railroads into the region, the development of chemical fertilizers that enabled the land to be productive, and the clearing of land by timber companies. Some settlers of the interior of the Coastal Plain were from out of state, but many were cotton farmers who moved from the worn-out Piedmont lands. Between 1870 and 1910, the population of the "Wiregrass Plain" increased 445 percent, the fastest growth of any section of the state in any forty-year period.[10] So by 1900 the state of Georgia was fully occupied.

Changing Agriculture

With the settlement of Georgia complete at the beginning of the twentieth century, the extent of land cultivation was near maximum. During the preceding century the acreage in clean-tilled crops had increased rather steadily and reached a peak between 1910 and 1920 when row crops occupied about nine million acres (Figure 2.1). This is certainly an underestimate of cleared land. In addition to orchards, pastures, and other minor enterprises, there were large expanses of idle or abandoned lands (also shown in Figure 2.1). Although idle land was not listed in earlier surveys, it is likely that there were two million acres or more because the census of 1924 lists 2.57 million acres in this category.[11] Diversification was nearly nonexistent at the time with cotton and corn accounting for nearly all of the cropped acreage. Although cotton and corn were almost equal in acreage early in the century, corn was not a cash crop. It was used mainly as feed for mules that pulled the implements to produce the cotton and for the few hogs, milk cows, and chickens kept on the farm. Some quantity of corn was also used as a staple food for the family. After an almost continuous depression since the civil war, the first two decades of the twentieth century were prosperous for Georgia farmers and high cotton prices encouraged full utilization of cleared lands.

The prosperous times were not to last, however and three blows to agriculture in the 1920s started the decrease in row-crop agriculture that lasted for the rest of the century. The price of cotton dropped precipitously in the early 1920s, the boll weevil invaded Georgia in the same decade, and the Great Depression at the end of the 1920s was the final straw. In 1919, the gross value of Georgia's farm production was six times higher than in 1899, but the price of cotton dropped from 35 cents per pound in 1919 to 16 cents in

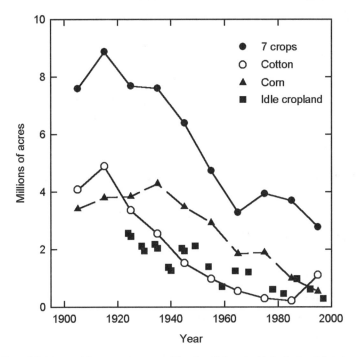

Figure 2.1. Acreages of the most common crops of Georgia and idle cropland during the twentieth century. The total for seven crops includes cotton, corn, soybean, peanut, sorghum, tobacco, and cowpea. From the Georgia Agricultural Statistics Service (1950–2000).

1920 and corn dropped from a dollar to 66 cents per bushel. To make matters worse, cost of production remained high. The state production of cotton dropped from the normal of 1,500,000 to 2,000,000 bales before 1920 to less than 600,000 bales in 1923.[12] Production in Greene County in the old Cotton Belt declined from 20,000 bales in 1919 to 333 bales in 1923. The state average yield from 1920 to 1924 was only 133 pounds per acre, the lowest five-year average since the 1870s. During the 1920–1925 period the farm population dropped by 375,000 and 3.5 million acres of land was removed from cultivation.[13] The last half of the 1920s was not as bad as the first and some of the cotton acreage was recovered; so the drop in the 1920s ten-year average in Figure 2.1 is less than 3.5 million. Finally, the Great Depression that began in 1929 worsened the plight of farmers and led to the New Deal programs of the Roosevelt administration. So the depression of agriculture after World War I lasted for most of the 1920s and agriculture did not fair much better in the 1930s.

Diversification of agriculture had been preached by many since before the civil war, but the economic and social traditions that made cotton culture attractive lasted until well into the twentieth century. The proportion of Georgia's cropland planted to cotton increased from about 30 percent just after the civil war to 47 percent in 1900.[14] However, the triple distress of unstable cotton prices, the boll weevil, and economic depression gave diversification a new emphasis and after the 1930s, cotton never again dominated Georgia and the South as it had for nearly a century and a half. Although cotton continued to account for nearly half of the row crops planted until 1930, by 1940 the figure had dropped to 24 percent. As shown in Figure 2.1 row crop agriculture decreased steadily from the 1920s to the 1960s, but the decrease is only an indication of the changes in the landscape during the period.

Georgia agriculture in the early 1900s was a subsistence culture. A census worker in 1910 figured that the average Georgia farm supplied for each of its occupants only 2/3 pint of milk per day, 2 eggs and 2/3 ounce of butter per week, and 1/3 of a hog, 1/12 of a beef carcass, and 1/100 of a sheep per year.[15] The farms were too many and too small. The number early in the twentieth century rose to over 300,000 and farm size reached its lowest at about 80 acres. In 1920, more than half the farms in the state were less than 50 acres in size and these farms cultivated an average of 21 or 22 acres.[16] The typical farm had 15–30 acres of cotton, one mule to cultivate the fields, one or two milk cows, and a few pigs. If the number of livestock in the 1930 census is divided by the number of farms, then the average farm had 1.5 mules or horses, 1.3 milk cows, 5.1 hogs, and 30 chickens. Of course, most farms had a mule or horse to cultivate cotton, but many had no other livestock. The average farm also planted 14 acres of corn. According to the 1930 census 67 percent of Georgia farms were classified as cotton farms,[17] but the average was only 20 acres of cotton per farm. Because all of the cotton had to be harvested by hand, this was about the maximum one family could cultivate.[18] So, in the 1930s Georgia farms were truly one-horse farms supplying a meager living for those on the land. They were "family farms," but definitely not the kind envisioned as ideal at the end of the twentieth century.

A major reason for the subsistence agriculture of the first half of the twentieth century is that over half of the farmers did not own the land they tended. Tenants operated 60–70 percent of the farms from 1900 to 1940. Tenancy posed two primary problems for prudent land use. First, the tenants had little incentive to conserve the soil, and according to Range (1954), there was little supervision of tenants by landowners. Second, tenants seldom stayed on the same farm long enough to start or maintain sound management or conservation measures. Although the percentage of farms operated by tenants had decreased by mid-century (to 42.8 percent), about half of the tenants stayed on the same farm no longer than one year.[19] It is hard to realize now, but tenants had little means to escape the system that held them at the poverty level. Most were not educated and there was little alternative early in the century to farming for a living. The income from the cotton crop was barely enough to survive and there was little chance to save enough to buy a farm.

It is little wonder then that the sharecroppers and small farmers at the end of the cotton era left the farm in droves. Although there is still today a view of the farm as a peaceful, clean, and healthy place to live, the reality of farm life and especially that of the first half of the twentieth century is quite different. Dr. Harold Nix of the University of Georgia said that the idyllic country life "is mostly an urban myth, created by people who are harried by city traffic but have never sweated to make crops grow or put in sixteen hours on a dilapidated tractor under a hot, South Georgia sun."[20] The prospect of a better life drove most of them to town. Even in the 1990s there was stress on Georgia's farms about the desirability of the farm life. Dodge County was taken as representative of Georgia's counties in a survey by Bartlett (1993) that showed the tensions between the attractiveness of country life and the hard work, long hours, and financial hardships.

The causes of an almost full century of tenant farming with its insidious effects on the land of Georgia are too many to be examined here, but lack of education and lack of means were two of the strongest. Even when the tenant-cotton agriculture ended and many of the former tenants moved to town, their lot was not improved as much as they hoped. Labor might have been easier and their pay better, but their health and education was not. The death rate in 1940 was 15.7 per 1,000 urban Georgians and 10.4 for rural residents.[21] As late as 1950, the median number of years of schooling received by rural farm people 25 years or older was only 7.1. Urban residents were not much better off at 8.8 years of school.[22] So the general population was not well educated at mid-century and earlier it was even less so. The low state of the rural population is dramatized, perhaps too much, in apparently true stories about two families in the early part of the twentieth century.

The first is about a Georgia family that spread over several counties.[23] The family name was descriptive of the lifestyle they endured. The "Bunglers" shifted from place to place and occupation to occupation with no change in the poor education, landlessness, and lack of socialization that was their lot. In one county, where more than a hundred Bunglers lived, a doctor who examined many of them found them heavily infested with hookworms. "One woman exhibited blisters on the soles of her feet which were caused by walking through the hot sand for a distance of four miles 'to big meetings' at her church." A farmer is quoted as saying that the Bunglers who lived on his farm as tenants spent at least one-fourth of the family income on snuff. The author was told by a family member that "the Bunglers are decreasing in every way except numbers."

The second story is told by Sisk (1975) of a family in the Sandy Creek area of Madison County. The Sandy Creek watershed was to become the first soil conservation demonstration project in Georgia.

> The farmer moved to his first 100 acres on Christmas Eve in 1916. There were nine children in the family ranging from two to thirteen years of age. In 1917, he worked 25 acres and made eight bales of cotton and ten loads of corn. By clearing some extra land in 1919 and working thirty acres, he managed to produce eleven bales of cotton and nine loads of corn.
>
> By 1925, yields had dropped to five bales of cotton and seven loads of corn and he was in debt $300, a nearly astronomical sum for those days and times. There were twelve children in the family then and the land was washed so badly that he had either to move or practically starve his family. He moved.
>
> But his sad experience followed him. (A common sight in those times was a wagon loaded with household goods followed by a lean milk cow tethered to the tail gate and a mangy hound

dog trotting along under the wagon.) By 1926, on the new place, he had made thirteen bales of cotton and six loads of corn and went into debt $200. He moved again to try another farm.

It is perhaps too depressing and too derogatory to say that these two families were typical of the sharecroppers of the early 1900s, but the level at which most of Georgia's rural people lived was not much higher. The effect of the agricultural system on its land resources, regardless of the complex causes and blame, largely justifies them as bunglers all.

Diversification

The number of farms and farmers has decreased steadily since 1920, and farms cultivated by tenants have almost disappeared (Figure 2.2). The reasons for these trends are many, and too complicated to discuss in detail, but some are rather obvious and relevant to the changes in land use. Increases in the uses of machinery, which accelerated after World War II, allowed one farmer to do the work of many laborers. The mechanical cotton picker meant that the size of a cotton farm was no longer limited by how much a family could pick by hand. Industrialization drew many farm laborers and sharecroppers from the farms to the cities. The use of fertilizers, pesticides, improved crop varieties, and other technology increased production and decreased the labor required to produce agricultural products.

In the twentieth century agricultural efficiency increased to an extent that released an overwhelming percentage of the population from the labor for sustenance. In antebellum days most people farmed and Georgia farmers probably fed 10 percent more people than those living on the farms. In 1920 each farm family probably fed an extra family of the same size.[24] Today less than 2 percent of Georgia's population lives on farms.[25] So the people not being required to produce food for themselves have moved off the farm and concentrated in the cities. As the number of people required for production of food has decreased drastically, so has the number of acres. The percentage of Georgia land in farms decreased from 69 percent in 1950[26] to 28 percent in 1997.[27]

The dominance of Georgia agriculture by cotton for more than 100 years was finally broken by mid-century (Figure 2.1). The 5.5 million acres that were taken out of row crops in the fifty-year period from 1910 to 1960 were diverted to several enterprises. Animal agriculture changed from a neglected and part-time enterprise to a dominant one. From World War I to 1940, the percentage of gross farm income from livestock and poultry was only 15–20 percent, with income from crops and government payments making up nearly all of the rest. In 1920, the cotton crop alone was valued at $229 million, twice that of all the livestock on Georgia farms,

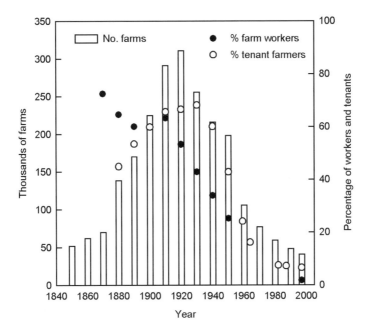

Figure 2.2. Changes in the number of farms, the percentage of people working on farms, and the percentage of farmers who were tenants since 1850. From Range (1954) and the US Bureau of the Census (1920–2000).

including mules which were the most valuable.[28] After mid-century, however, income from livestock and poultry exceeded cotton. By the late 1980s, livestock and poultry accounted for over 50 percent of cash sales, with poultry bringing in the most cash. Crops, including horticultural, made up only 35–40 percent of gross income, and cotton only 2–3 percent.

The downward trend in row crops continued throughout the twentieth century. The State Soil and Water Conservation Commission (1995) estimated that there was a decrease of 1.39 million acres of cropland from 1982 to 1992. However, the decrease in row-crop acreage during the twentieth century was not uniform across Georgia. In Figure 2.3, harvested acreage for 1935 and 1997 is shown for eight selected counties in the Mountains, Piedmont, interior Coastal Plain, and the flatwoods of Southeast Georgia. The

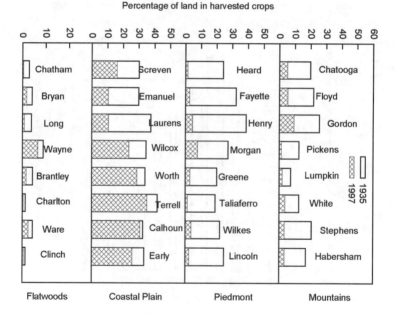

Figure 2.3. The change in acreage of harvested crops from 1935 to 1997 in selected counties of four regions of Georgia. Data from the US Bureau of the Census (1920–2000).

decrease in acreage in the northern half of Georgia was large, totaling 74 percent for the eight mountain counties and 86 percent for the Piedmont. Even in the Coastal Plain where agriculture is now most concentrated, row crops decreased. The drop in harvested acreage was greatest in the eastern Coastal Plain, averaging over 60 percent in Laurens, Emanuel, and Screven Counties. There was a decrease in harvested acreage in counties along the coast as well, but row-crop agriculture never accounted for much of their land.

In addition to the decrease in the acreage of row crops, the landscape patterns changed. Farms became fewer and larger. In 1930, Georgia had over 250,000 farms (Figure 2.2) and 75 percent of them were less than 100 acres. At that time, the Georgia landscape was very fragmented with many, small, irregularly-shaped tracts of open and forested land. These small farms each had a variety of land uses, small fields of cotton, corn, peanuts, pasture, some woodland, a vegetable garden, and the farmstead. Most of the more valuable crops were in fields that were fenced. It seems unbelievable in an era of modern supermarkets, but in 1939, there were nearly 200,000 acres of home vegetable gardens.[29] This acreage was nearly equal to that of cotton for the whole state from 1978 to 1984.

Many of Georgia's small farms had idle land "laying out"; in the 1930s, there were about two million of these idle acres (Figure 2.1). The briar patch, favored home of Brer Rabbit in the Uncle Remus tales of Joel Chandler Harris, was a prominent feature of these idle acres early in the twentieth century. Blackberry bushes are the most common briar of abandoned Georgia croplands until trees take over. They are a favorite haunt of not only Brer Rabbit, but many other small animals, birds, and reptiles of open fields and forest edges. Quail hunters know that when a covey is flushed, the briar patch is a likely refuge for the scattered, single birds, and the "real" bird hunters wear thick heavy trousers or chaps to protect themselves from the thorns. They also know that idle cropland with thick, bushy fencerows are good places to hunt. Two million idle acres in the 1920s and 1930s with an abundance of weed seeds, and woods that were frequently burned, made habitat ideal for quail. These idle acres, and fencerows thick with bushes and briars for cover on nearly every farm, made quail the favorite game bird and rabbit the most hunted animal. The disappearance of so much of this habitat on larger more mechanized farms is likely the main reason that the populations of both quail and quail hunters have diminished.

Today the rural landscape is made up of larger tracts, more regular in shape.[30] A survey of twenty farms in twenty counties about 1930 showed that the average field in Georgia consisted of only 5.6 acres.[31] Fields are now larger and unfenced to accommodate large machinery. In 1994, there were only 43,000 farms (Figure 2.2) and they averaged about three times larger than those in 1930. Edges between land uses and dissection of the landscape decreased from the 1930s to the 1980s.[32] Forest tracts are larger, more thickly stocked with trees, and in most cases, continuous for miles. Brender (1952) describes an area in the lower Piedmont previously in cultivation as follows, "The young pine forest begins virtually at the city limits of Macon and rolls away in ten, 20, and even 30-mile unbroken tracts."

Much of the land taken out of row-crop cultivation was put into pasture. The trend of pasture acreage during the century is not easy to trace because pastures have been classified differently in the surveys. In surveys early in the century such classifications as "grazed woodland" and "plowable pasture" have confused the real uses of the land. Much of the grazed land early in the twentieth century was not what we call pastures today. Before 1940, a great majority of the counties south of the fall line had "no fence laws," meaning that cattle were not confined, but free to roam any land that was not fenced-in to protect crops. In spite of the confusing classification, pasture land has at least doubled since the 1940s. In the early 1940s, there were 1.5 million acres of "plowable pasture,"[33] and in 1989–1990 it was estimated from satellite imaging that there are 3.34 million acres of pasture.[34]

The twin goals of diversification and larger scale that many had preached for successful farming became a reality, but there were drawbacks to the trends. The exodus of people complicated rural living. Small towns dried up. Community schools closed in a wave of consolidation and children were bused to larger towns. Country stores that had been gathering places for country shopping and socializing gave way to convenience stores in small towns and malls in the larger ones. The mainstay of small Georgia towns early in the century was the cotton gin. Over the years 1936–1939 there were 1,560 in the state.[35] At the end of World War II, there were still 1,149 cotton gins in the Piedmont of the southeast, but by 1967 there were only 138 active gins.[36] By 1984, there were only fifty-three active gins left in Georgia.[37]

Like many others, Bill Winn (1968) lamented the shifting of the population. He said, "With the people gone, the countryside of Georgia has taken on an oppressive atmosphere." He quoted a one-time South Georgia farmer as saying about the emptying of the countryside, "It's the new machinery done it. It's the machinery got rid of the small farmer, and it's the chemicals taken the hoe-help away." Whether mechanization and other technologies caused or resulted from movement of people from the country (it probably was a combination of cause and result), the shift in population had a profound effect on the environment. For the countryside, the effect was positive. After all, the main root cause of most pollution is concentration of people. Most of the land deserted by crops and people was taken up by forest.

Reforestation

The extent and condition of forests is regarded by many as an indicator of the health of the environment. In Georgia, the extent of forest has changed drastically since colonial days. The Georgia

Forestry Commission[38] issued a publication entitled "Georgia's Fourth Forest" to describe forest conditions late in the twentieth century. This title arises from the concept that Georgia's forest has been cut down and regenerated repeatedly. The first forest was, of course, the one cut down by settlers for cultivation of crops and other uses. The second forest began by natural regeneration when the cropland was abandoned, or when in some parts of the Coastal Plain and Mountain regions cut-over timberland was left to natural regrowth. This second forest was begun at different times depending on the time of settlement and how long the land was cultivated before it reverted to forest. Spillers and Eldredge (1943) estimated that about 8.8 million acres of second-growth forest had produced saw-timber size logs by the mid-1930s, implying that the regeneration had begun near the turn of the century as was also stated in "Georgia's Fourth Forest." The 'third forest' was the current one at the time of publication, and the "fourth forest" was being established late in the century.

The change in extent of Georgia's forest can only be surmised from narrative accounts prior to the twentieth century and not with much accuracy even in the early 1900s. Surveys by the US Forest Service began in the early 1930s (Figure 2.4). The acreage increased from about 20 million in the 1930s to nearly 26 million acres in the early 1960s. The acreage is likely to have been considerably lower before the 1930s because census figures for farm forest (that forest on land classified as farms) show an increase from the early 1920s to mid-century (Figure 2.4). The jump in acreage from 1958 to 1963 was partly due to the Soil Bank Program, which paid farmers to take land out of row-crop cultivation. The Soil Bank Program was responsible for planting of about 700,000 acres of trees in Georgia in the years from 1956–1960.[39] There followed a period of increased agricultural production spurred by high prices when acreage equivalent to the Soil Bank acres was cleared of forest and cropped. In spite of a decrease of about 10 percent from 1963 to the

1990s, there was still 18 percent more forest in Georgia in 1997 than in the mid-1930s.

The extent of forest in the 1930s may have been overstated by the classification of forest land. Many tracts that were little more than abandoned fields that had partially grown up with bushes and small trees were probably classified as forest. Of the 21 million acres of forest listed by the Citizens Fact-Finding Movement of Georgia (1946) in 1938, over three million acres were classified as "poor to nonrestocking area." The first forest surveys in the 1930s found that in forests of the five Forest Service Districts of Georgia the average amount of wood ranged only from about 6 to 11 cords per acre. The average in the 1990s was nearly 20 cords per acre. Spillers and Eldredge (1943) said about Georgia's forest in 1935, "Most of the

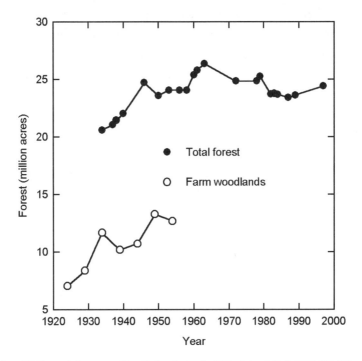

Figure 2.4. Changes in the acreage of forest in Georgia since the 1920s. Farm woodland acreage data from Saunders (1960); total forest acreage from various sources.

trees are small. Although the sites usually are good, the stands are less than half stocked owing to overcutting and to the frequent occurrence of forest fires...." Their remark about small trees is supported by surveys that showed 75 percent of the trees in the northern two-thirds of the state were less than 5 inches in diameter.[40] Although trees in the same survey areas were skewed even more toward small sizes in 1997 – (80 percent below 5 inches) – the stocking was two and a half times more, an average of 720 trees per acre in 1997 compared to 290 in 1935. The poor stocking and poor condition of the trees was described by Spillers and Eldredge (1943) as follows, "As to quality, much of the pine volume is in trees that are limby and rough, owing to the open condition of many of the stands, especially those in old fields." There was in the 1930s more than 10 percent of the forest land in Central and North-Central Georgia that was still eroding to some degree. Land with well-developed forest is not subject to erosion. So, although more than half of Georgia's land was covered by forest in the 1930s, it was certainly not comparable to the forest before settlement of the state, nor comparable to the forest of today.

Replanting has been a hallmark of forest management in Georgia in the last half of the twentieth century. Although in the early days forests regenerated slowly from old fields and cut-over timberlands, planting of forests became widespread about mid-century. As shown in Figure 2.5, the number of acres replanted in the 1930s was near zero, but climbed steadily since the mid-1940s, so that in the 1990s, 300,000 to 400,000 acres were being replanted annually. Two peaks are evident in Figure 2.5 that show the effects of two government programs on tree planting. The first was the Soil Bank Program in the late 1950s and early 1960s. In some counties, especially in the Piedmont, the effects of the Soil Bank Program were dramatic. In 1955, the year the soil bank started, one-third of the land in Madison County was idle cropland. Ninety percent of this land had been planted to trees in the Soil Bank Program by

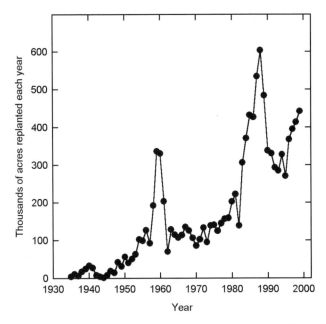

Figure 2.5. Acreage of forest replanted since the mid-1930s. The 1934–1985 acreage from Rivers and Loggins (1989), 1986–1999 acreage from Georgia Statistical Abstract (Various years).

1960.[41] The second peak in Figure 2.5 resulted from the Conservation Reserve Program begun in the 1980s. About 600,000 acres of trees were planted in 1988, the year of most sign-ups. Tree planting in the depression years by the Civilian Conservation Corps is hardly visible in Figure 2.5. Drives by environmental and civic groups late in the twentieth century to improve the environment by planting trees are laudable and helpful, but they pale in comparison to the unheralded millions of trees that farmers and foresters have planted in the last half of the twentieth century.

Claims of recent changes in Georgia's forests are often contradictory. Harris and Frederick (1991) concluded that Georgia's forest was being lost at 0.4 percent per year. McCrary and Kundell (1997) reported forest acreage decreased from 22,096,200 in 1982 to 21,996,400 in 1992, a decrease of 4.5 percent. Sheffield and Johnson (1993) gave the acreage in 1982 as 23,733,700. In 1999,

Bob Lazenby, deputy director of the Georgia Forestry Commission in Macon was quoted as saying, "With an average of 613,090 trees planted each day for the past 20 years, and millions more regenerated naturally, Georgia landowners are growing substantially more trees than they are cutting." He also said that Georgia's forests had increased by 441,000 acres from 1989 to 1997.[42] As can be seen in Figure 2.4, however, although the acreage has gone up and down, and peaked about 1963, it has not changed more than 10 percent since the middle of the twentieth century.

In addition to the replanting of lands to trees, better management of woodlands has resulted in much greater production of wood from forests. The yield of lumber from Georgia's forest was from 1.5 to 2.0 billion board feet per year from 1940 to about the mid 1970s, but since that time it has risen to about 3.0 billion board feet (Figure 2.6). The harvest of pulpwood, usually smaller trees used mostly for paper products, has increased from less than one million cords in 1940 to more than 10 million cords in the 1990s. The increased harvesting of the forest has resulted from increased productivity rather than depletion of the growing stock. In fact, the standing inventory of wood in Georgia's forest has increased from about 200 or 250 million cords (saw timber and pulpwood) before the 1950s to about 450 million cords in the 1990s. This is an average of about 19 cords per acre in the 1990s compared to less than 10 cords per acre in the mid-1930s. This phenomenal increase in forest productivity over the last seventy years is a result of more forest land, but mainly wiser management. Better management is the reason for the 140 percent increase in forest productivity per acre from 1935 to 1982 cited by Turner (1987). It certainly reflects an improvement, not a degradation, of the environment of Georgia.

In spite of the facts that show there have been increases in both harvest rates and inventory of Georgia forests, there are some claims of decreased forest growth. Zahner et al (1989) measured the tree rings of loblolly pine in the southern Piedmont and concluded that there was a decline in growth, which averaged 1 percent per year since 1950. Likewise, Sheffield et al (1985) found that radial growth of trunks of loblolly pine was 20 to 30 percent less from 1972 to 1982 compared to 1961 to 1972. In both cases the decreasing trend was attributed in part to "atmospheric deposition," and Zahner et al (1989) concluded that "The regional trends in levels of ozone and other possible pollutants indicate that they should be seriously considered as correlative agents." However, other factors such as drought, disease, increased crowding, and age could not be ruled out. In contradiction to this trend of decreasing growth, West et al (1993) found about 40 percent greater growth in diameter of trunks of an old-growth stand (100 to 400 years) of long-leaf pine in Thomas County from about 1950 to the early 1990s. They explained that "The increased growth cannot be explained by dis-

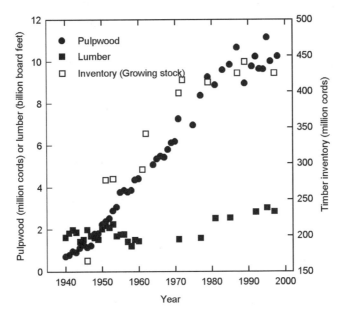

Figure 2.6. Changes in annual production of pulpwood and lumber and in the inventory of Georgia's forests. From Spillers and Eldredge (1943), Forest Survey Releases of the Southern Forest Experiment Station, and USDA Forest Service, Southern Research Station Resource Bulletins SRS-14, 38 (Up to 1978, pulpwood does not include chipped pulp).

turbance, stand history, or trends in precipitation, temperature, or Palmer drought severity index over the last 57 years. Increased atmospheric CO_2 is a possible explanation for initiation of the observed trend, while SO_x and NO_x [SO_x is sulfur oxides and NO_x is nitrogen oxides] may be augmenting this phenomenon." So pollution has been used to explain both a decrease and an increase in growth of pine forest in Georgia.

It is likely, though, that changes in pollution have played a minor role if any, and that the increased productivity of Georgia forests has resulted from better management. The trend in Georgia is atypical for the United States only by the degree of improvement. According to Moffat (1998), national forest growth has outpaced forest clearing over the last fifty years, so that the volume of timber in the nation has increased by 30 percent. The south did much better; wood volume "...went up by 70%...where softwood plantations now thrive on soils degraded by decades of cotton farming...." As indicated above, Georgia's wood inventory increased by nearly 100 percent during that period (Figure 2.6). Not only are most forests now "fully stocked," but much of the state's forest is managed as a crop. Silvicultural practices, such as replanting, land preparation, weed control, fire protection, thinning, and removal of diseased or cull trees are common.

The increased productivity of Georgia's forests is due in large measure to control of fire, which was a common occurrence in the early twentieth century and primarily intentional, not accidental. Fire was often used to improve grazing in woodland "pastures" which furnished much of the forage for livestock early in the century. While frequent burning improves the quality of woodland pastures, it also prevents establishment of young trees to restock the forest. Spillers and Eldredge (1943) related that in the 1930s "only 27 percent of the forest area receives the benefits of organized forest-fire protection." In the 1930s and probably earlier, four to five million acres of forestland were burned annually (Figure 2.7).[43]

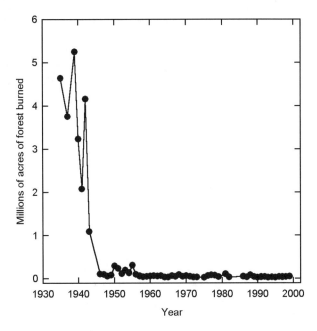

Figure 2.7. Changes in the acreage of forest burned by wildfire in Georgia since the 1930s. From Spillers and Eldredge (1943) and the Georgia Forestry Commission (1946–1999).

Beginning in the 1920s, campaigns against wildfire began to raise the awareness of farmers and timber men to the damage wildfires cause. Dr. Charles Herty, the man most responsible for the beginning of the pulp wood industry, was a prime mover in the education about the detrimental effects of fire in timberlands.[44] In 1928, a group called the "Dixie Crusaders" was formed by the Southern Forestry Congress, the American Forestry Association, and state foresters to inform country people in Mississippi, Georgia, and Florida of the importance of forest resources and the damage caused by forest fires. The Crusaders traveled throughout the three states lecturing and showing two movies, *Pardners* and *Danny Boom*. They distributed posters, hand bills, and book covers to almost two million parents and school children.[45] These early efforts and the introduction of Smokey the Bear by the US Forest Service deserve much of the credit for decreasing wildfires in Georgia by 95 percent by the 1950s (Figure 2.7).

Because frequent wildfires prevented regeneration of forests, foresters began early in the century to discourage burning. The campaign reached its peak in the 1930s and 1940s, and wildfires have been a minor problem since then. In the latter part of the twentieth century, the benefits of burning for a variety of purposes was recognized. According to Wade and Lunsford (1989), controlled burning is used to reduce accumulation of combustible materials, dispose of logging debris, regenerate Southern pines, improve wildlife habitat, control insects and diseases, improve access for working in the forest, increase range (grazing) habitat, and increase fire-adapted species. The recognition of so many benefits of burning has led to an increased use of fire in forest management. It has not, however, resulted in an increase in wildfires (Figure 2.7).

The elimination of fire had far-ranging effects on Georgia's forests and not all of them were desirable. Because many hardwood species are more susceptible to damage by fire than pines, elimination of fire causes invasion of pine forest by hardwood species. In fact, in the Piedmont in recent years there has been the observation that the balance between hardwoods and pines in mixed forests is shifting towards more hardwoods. Georgia's forests may have fewer fire-resistant species today than the forest of 200 years ago. The species of trees in a 888 square mile area of the Piedmont just above the Fall Line near the Oconee River were counted in 1990 and compared to species listed on land lines in the survey for the Georgia land lottery in 1801–1804.[46] It was found that the species intolerant of fire are more prevalent today than in the 1801–1804 surveys. The conclusion was that reduction of forest fires by fields and roads that prevented spread of fire and by organized fire control in the twentieth century had changed the composition of forests from that of pioneer days when fire must have been a fairly common occurrence in the forest.

It is common practice to remove hardwoods from pine forest because pine is, in most cases, more valuable and faster-growing than hardwoods. In some cases, hardwoods are removed by herbicides, but it is common to use "prescribed" or "controlled" burning to eliminate the hardwoods when they are small. Controlled burning is ten to twenty times cheaper than use of chemical or mechanical means of reducing hardwood competition.[47] Prescribed burning is also used to make the woods more favorable habitat for some species of wildlife.[48] Controlled burning is accomplished by plowing a "fire-break" around the area to be burned and setting fire to the area. Burning is done only when atmospheric conditions are favorable for containing the fire. Still days are selected when the woods are not too dry, and preferably when the humidity is high. Such burning is coordinated with foresters of the Georgia Forestry Commission and monitored to prevent fire from escaping. Although the decrease in wildfires has been a bonanza for forest production in Georgia the negative effects on some species of wildlife have been substantial.[49]

Wetlands

Reports about the benefits of wetlands and alarm about losses of wetland acreage have been common in the last half century. This is a complete turnaround in public opinion. The wetlands of today were the repulsive swamps of earlier centuries, and even the early part of the twentieth century. Swamps were undesirable either because they were feared to be the source of sickness, they were feared to harbor dangerous or mysterious creatures, or they were useless for agriculture. Drainage of these swamps was perceived before the middle of the twentieth century to be an improvement of the environment. Barrows et al (1917), in a book entitled *Agricultural Drainage in Georgia* said about drainage of a swamp in Gwinnett County, "A notable result of the improvement [ditching the channel] of Big Haynes Creek has been the elimination of the malarial conditions that formerly existed along the creek." The Georgia Drainage Law of 1911 required that for land to be included in drainage projects for government assistance, the proposal "also must set forth that the land is too wet for cultivation and that the public health and welfare will be promoted by a system of drainage."

Georgia ranks fourth among the lower forty-eight states in total wetland acreage with about 7.5 million acres [50] (but see below the confusion about acreage). The extent of wetlands is small in the northern portion of the state, but large in the southern part. According to the Georgia Water Use and Conservation Committee (1955), "About 6,000,000 acres, or approximately 1/6th of the state, is seasonally flooded, waterlogged or permanently covered with water. Wooded swamps make up more than half of this acreage and include some of the best hardwood timber and waterfowl production areas in the state." Although the term "wetland" was not in wide usage in mid-century, it is obvious that the state was beginning to value its former "swamps." Even if its large acreage of wetlands has been diminished by drainage, Georgia has lost a smaller percentage of its original wetlands than any other southeastern state.[51]

The trends in wetlands of Georgia are bound up in the efforts to drain swampy land since the settling of the colony. Although swamps were considered to be sources of disease, most drainage was for agriculture. One of the first crops to be cultivated in Georgia was rice along the coast. Wet areas had to be drained to prepare the land for seeding and harvest of the crop. These areas had to lay low enough that they could be flooded from nearby streams. So conversion of wetlands began early, but conversion was not limited to colonial days nor to the areas near the coast. There was a push for drainage of agricultural lands that coincided with the prosperous first two decades of the twentieth century. The Georgia Geological Survey published surveys of land needing drainage in 1911 and

1917[52] and many drainage districts were established. As late as 1955 it was considered that "The need for work in drainage is widespread over the state. North of the fall line there are many areas of formerly good bottom lands which are now too wet for any cultivation. Often they have passed from good production to swamp in as little as twenty-five years."[53] A "needs inventory" of Georgia watersheds in 1958 cited 6,694,500 acres as having "the problem" of poor drainage (includes both drained and undrained lands) and 4,102,900 acres as needing "project action." The acreage needing "project action" was not all that needed drainage, but was acreage "that cannot be adequately protected or treated by individuals or groups without the assistance of organized groups such as those authorized by Public Law 566."[54]

Except for rice culture on the coast, early efforts to drain land was in the Piedmont where, as indicated above and in chapter 3, bottomlands had become choked with sediment washed from the uplands. The first drainage district organized under the drainage law of 1911 was on Big Haynes Creek in Gwinnett and Walton Counties.[55] By 1917, forty drainage districts had been formed or planned, but only three were in the Coastal Plain. Apparently the Coastal Plain counties were slow to form drainage districts because of the large projects needed and the high cost associated with large projects. The Bureau of the Census listed 46,592 acres as having been drained in 1930 and 80,514 acres in 1940. The eleven counties having enough drained acreage to be listed separately were in the Piedmont and they accounted for 60 to 70 percent of the total. One single county, Franklin, was listed as having drained 23,619 acres by 1940. The county with the next most drainage was Jackson with only 5,194 acres. The above numbers apply only to land drained by organized drainage districts, because the first Bureau of the Census estimate in 1920 listed 274,688 acres as drained.

But the vast majority of wetlands are south of the fall line, especially in the counties along the coast. The largest freshwater wetland in the state is the Okefenokee Swamp near Waycross. There was very early a plan to drain this swamp to harvest its timber and cultivate it. A survey ten years before the Civil War concluded that, "there would be no engineering difficulty in draining the whole swamp perfectly and rendering available the enormous amount of cypress timber as well as thousands of tons of muck, which, with the aid of the Satilla marls, would convert the sandy as well as the red clay lands in the border into market gardens."[56] There was a canal dug 45 feet wide, 6 feet deep, and about 12 miles long which was used to transport timber out of the swamp, but which apparently never was completed enough to drain the swamp. The desire to drain the swamp changed to a desire to prevent its drainage. There was enough fear about too much natural drainage and resulting fires during dry years in the middle of the twentieth century that a dam or "sill" was constructed to reduce water flow out of the swamp.[57]

In the latter half of the twentieth century there has been a decrease in the efforts to drain swamps and an increase in the efforts to save them as wetlands. Much of the increased awareness about the benefits of wetlands in Georgia has come from the writings of Wharton (1970, 1977) who was a leader in saving the Alcovy River swamp from drainage. The benefits of wetlands are that they modulate the flow of streams so that flood and drought are reduced. In addition, they filter pollutants from the water and provide habitat to many creatures that require the shallow, slow-moving water and intermittent flooding for their life cycles. Because of the increased awareness of the benefits of wetlands, there are now restrictions on the drainage of wetlands for agriculture or other purposes. Although channeling of streams and preservation of wetlands are opposite management schemes that have other objectives, flood prevention was claimed for both. The argument in favor of wetlands is that the large flat areas along streams hold large volumes of water and slow the movement so that the water does not rush down the valleys and inundate residential or other important areas. On the other hand, the argument of the drainage engineers was that larger stream channels carried more water and carried it more quickly from flood-prone areas reducing the flooding danger.

In spite of the recent efforts to preserve wetlands and in spite of the claims that they are being destroyed at alarming rates, the facts behind the trends for Georgia wetlands are as much of a "bog" as the wetlands themselves. The counting of wetland acres is imprecise, to say the least. Estimates made in Georgia since the early 1960s vary from 3,513,789 acres[58] to 10,712,661 acres[59] a difference of over 200 percent (Table 2.1). The differences are because of the variable definition of wetlands and difficulties in estimating them. Many lands are wet some of the year and dry at other times and may or may not be classified as wetlands. Some estimates are made from

Table 2.1 Estimates of Georgia wetlands during the 20th century.

Acres of wetlands	Source and notes
7,919,469	Barrows et al., 1917. "In need of drainage"
5,919,500 (mid-1950s)	Shaw and Fredine, 1956, Nat. Wetlands Inventory
10,712,661 (1961)	U.S. Department of Agriculture, 1961a[1]
6,694,500 (1958)	Ga. Soil & Water Cons. Needs Inventory Comm., 1962[1]
4,143,000	State Soil & Water Cons. Comm., 1981 (SCS 1979 est.)
5,444,597 (mid-1950s)	Kundell and Woolf, 1986, rev. Wetlands Inventory[2]
5,298,000 (mid-1970s)	Kundell and Woolf, 1986, from Hefner and Brown, 1984
3,513,789	Kundell and Woolf, 1986, from GDNR and Ga Tech study
4,831,300	Kundell and Woolf, 1986, Soil Conserv. Serv. study[3]
7,496,100	Ga. Stat. Abstr., 1998, SCS Nat. Resources Inv. 1992
7,792,000 (mid-1970s)	Hefner et al., 1994, National Wetlands Inventory[4]
7,714,000 (mid-1980s)	Hefner et al., 1994, National Wetlands Inventory
6,956,000	USDA, 1997, Agricultural Handbook No. 712
4,929,300	Ga. Dept. Nat. Resources, 1997. Satellite imagery

[1] Includes both drained and un-drained wetlands.
[2] This number was revised from Shaw and Fredine, 1956.
[3] Does not include Federal Lands and open water.
[4] Revision from Hefner and Brown, 1984.

aerial photographs and others from satellite images, and wetlands may be poorly visible or poorly delineated in such images.

The difficulty in counting wetland acres makes estimates of losses in the twentieth century questionable. It is rather widely quoted that Georgia has lost 20–25 percent of its wetlands since the late 1700s.[60] The estimate for the 1780s given in the US Department of Agriculture Handbook 712 (1997) is 6,843,000 acres. Compare this estimate to those given below. Kundell and Woolf (1986) quoted estimates of the US Fish and Wildlife Service that between the mid-1950s and the mid-1970s, 146,597 acres or 2.7 percent of Georgia wetlands were lost. However, Kundell and Woolf hedged about the precision of the estimates in the 1950s by saying, "Although these figures are considered statistically reliable at the national level, they may be less reliable at the state level." The state of Georgia recently put it more directly, "However, we still lack accurate assessments for current and historic wetland acreages."[61] The decrease from the 1950s to the 1970s was based on estimates of 5,444,597 acres in the mid-1950s and 5,298,000 in the mid-1970s. However, the number for the 1950s was revised downward from the original estimate of 5,919,500 acres (Table 2.1). Later, the acreage for the mid-1970s was revised upwards by 47 percent from 5,298,000 to 7,792,000 acres, apparently because of the poor quality of the aerial photographs used in the original estimates.[62] Using this revised acreage for the mid-1970s, Hefner et al (1994) estimated that Georgia wetlands were reduced further by 78,000 acres (1 percent) by the mid-1980s to 7,714,000 acres. This confusion of numbers is with estimates made by the same agency (US Fish and Wildlife Service) over the years, so it is not surprising that estimates made by others using different methods and definitions are even more variable (Table 2.1). It is surprising, though, how close the most recent estimates are to the 7,919,469 acres estimated by Barrows et al (1917) early in the twentieth century as needing drainage (Table 2.1).

The loss of wetlands has been estimated mostly by tallying those acres defined and identified as wetlands and subtracting the later numbers from the earlier ones. Losses could also be estimated by using records of acreage drained. However, information on the acreage of lands drained is even more unsatisfactory than the counting of wetland acres. For one thing, land once drained may be swamped again later if drainage ditches are filled with sediment. The US Bureau of the Census has reported acreage of drained lands over the years, but the numbers are suspect, as the bureau admits. The following justification for a change of methods in 1978 illustrates.[63]

> Results from the censuses of agriculture in 1920, 1930, 1969, and 1974, show that attempts to enumerate acreage of artificially drained agricultural land by individual farm operators were unsatisfactory. Current operators often have limited knowledge of functioning drains installed in the past. Operators often had difficulty in estimating the effectiveness of old drainage systems. The result was inconsistencies between censuses and gross underreporting of artificially drained land in the census of agriculture.

The earlier censuses that the bureau judged faulty reported 274,688 acres of previously drained Georgia farmland in 1920 and 193,486 in 1974. An example of discrepancy between the two agencies is an estimate of 94,728 acres from the Bureau of Census in 1950 and 149,914 acres from the USDA Soil Conservation Service (SCS) report for the same year.[64] For 1978, the Bureau of the Census (1981) used estimates of the SCS "based on their own records and contacts from other organizations or knowledgeable individuals." The estimate in 1978 of drained land in Georgia was 1,530,980 acres, ten times the acreage in the SCS report of 1950 and five times higher than Census Bureau estimates in 1961. So from Census Bureau drainage figures, it can be calculated that 2.4

percent of the wetlands tallied by Barrows et al in the 1910s had been drained by 1974. But using the SCS estimates for the 1978 census, the estimate of wetlands drained is 19.3 percent.

A striking example of the difficulty of enumerating wetlands is in the successive censuses and wetlands estimates in Franklin County. In the censuses from 1940 through 1960, Franklin County was listed as having about 23,000 acres of drained land covering about 13 percent of the county. Much of the drainage must have occurred during the second decade of the century because Geldert (1916) said of the drainage project already underway, mostly in Franklin, but also including Stephens County, "It includes 20,000 acres on which bonds will be placed." The census of 1920 shows 7,869 acres drained in Franklin County. In the 1974 census there were only 450 acres indicated as drained, and by the new method of the 1978 census, 3,750. The SCS, the source of the acreage in the 1978 census, had estimated in 1961 that only 210 acres had been drained.[65] That same SCS survey lists 14,732 acres in Franklin County as undrained wetlands, Kundell and Woolf (1986) list it as having no wetlands, and the Georgia Department of Natural Resources (1996a) estimated it had 3,422 acres. Possibilities for the large changes in the wetlands of Franklin County during the twentieth century are that drained land became swamped after the 1920s, land was drained that didn't need it, and/or estimates of wetlands and drainage were grossly in error. Franklin County is likely an extreme example of the errors in counting of wetlands, but it illustrates the imprecision in estimates of both the extent and loss of wetlands. With such dramatic differences in estimates, specific claims of wetland losses over the years are dubious for counties and for the state.

It is possible that there are more wetlands in Georgia today than when Europeans settled here. The US Department of Agriculture Handbook 712 (1997) gives the estimate for the late 1700s as 6,843,000 acres. In that same publication the acreage for 1992 is

6,956,000, representing not a decrease, but an increase of 1.6 percent over that 200-year period. Gains in wetlands occurred in seven other states during the period and it was explained in the handbook that the gains may be "due to underestimates of original wetlands, or represent real gains through incidental or intentional wetland creation or restoration associated with water impoundments and other projects." Acreage impounded (covered by lakes and ponds) has certainly increased greatly during the twentieth century. In the 1990s there were more than 500,000 acres in lakes and ponds of Georgia that did not exist in 1900.[66] Some of the increase in wetlands has been "natural." Areas drained earlier became swamped again.[67] The beaver, which nearly disappeared in the first half of the twentieth century, has since dammed streams all over the state. Wharton (1977) quoted Godbee and Price (1975) as claiming that "Between 1967 and 1975, beaver damage increased 128%, with 287,700 Georgia acres inundated." If this claim is true, beavers likely created more wetlands than were estimated as lost from the 1950s to the mid-1980s.[68] The efforts to save wetlands are usually justified by citations of their benefits and also on the claims of their rapid destruction. Acreage estimates over the years are too variable to support claims of rapid loss. Whether there has been a net decrease in wetlands since the settlement of Georgia, it is not likely that there has been a net loss since the 1950s, when the acreage of lakes and ponds, including beaver ponds, really began to increase. It is more likely that wetlands have increased.

Urbanization

A major change in Georgia that had an effect on many problems of the state, including the environment, was the large increase in population and movement of the population from the countryside to town. When the era of the tenant farmer ended, and especially after World War II, many of the former inhabitants of the small farms moved to the cities and towns. Actually the population figures do not show a mass migration to town, because the rural population did not decrease much from 1920, when there were 2.1 million rural people (Figure 2.8) and 310,000 farms, to 1960 when there were still 1.98 million. Rather, the one million people Georgia gained during that period were nearly all urban and they increased the urban population by more than 2.5 times. As shown in Figure 2.8, the population of Georgia is represented by an ever-upward curve, except for a slight slow-down in the 1920s. The population has more than doubled in the last half of the twentieth century, and despite only a modest decrease of the rural population they have constituted an ever-smaller percentage of the total. That proportion of the population decreased from about 85 percent in 1900 to 25 percent in 1990.

The nearly level rural population from the 1920s to the 1960s (Figure 2.8) is misleading because many of those classified as rural

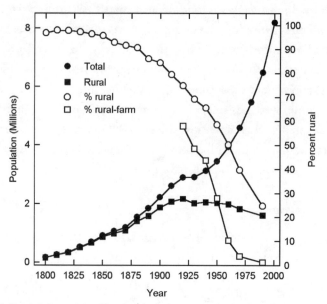

Figure 2.8. Changes in the population of Georgia during the last two centuries. The slope of the percent-rural line is affected after 1950 by a slight change of the definition of "urban areas." From US Bureau of the Census (1920–2000).

were actually suburban to the towns and cities. It is even more misleading if the rural population is confused with the people who make their living from farming. Many moved nearer to the small towns and large cities without actually being counted as urban. The growth of the large cities and loss of population by rural counties is only part of the picture. Even in the rural counties there has been a movement to town. Dodge County in the upper Coastal Plain south of Macon is fairly representative of rural counties in that it lost 21 percent of its total population and 37 percent of its rural population between 1930 and 1980. However, it lost 91 percent of its farm population and its urban population, although still small, grew by 74 percent.[69] In Georgia, even though a sizable portion of the population still lives in rural areas, less than 2 percent live on farms (Figure 2.8).

Not only has the population shifted to urban and suburban neighborhoods where opportunities and quality of life seemed better, but the countryside has become a better place to live. Instead of the models of clean, healthy country living that are often remembered or expected of the long ago, Georgia homesteads of the early 1900s were by today's standards unhealthy. The typical farmstead in the 1930s and 1940s was designed for servicing the farm more than for family living. Pigpens, chicken houses, cow lots, and mule barns were characteristic of most country homesteads, and were close by the house. The toilet was a wooden hut close enough to the house to be convenient in cold weather, but far enough away to minimize the odors. Farmyard smells were not as pleasant as in children's storybooks or in the colored memories of nostalgic elders. In many rural homesteads, chickens, dogs, and other animals ranged freely. There was little concern for the dropping of animal waste, except the unpleasantness of stepping in it. It is little wonder that internal parasites like hookworm and other stomach worms were the common cause of illness in barefoot rural children (Fite, 1984). There were also other maladies that troubled rather than threatened barefoot children of early twentieth-century Georgia. "Ground itch" and "Tetter worms" were fungal parasites that annually brought on "cures," such as the application of fresh black walnut juice that stained the feet black, and fresh cow manure that stained the psyche, whether or not it cured the feet. Although credit for the increasing stature of children and adults during the latter half of the twentieth century is often given to better nutrition, the freedom from parasites probably plays as large a role.

Houses of the early twentieth century were poorly designed to keep out the heat and cold. As late as the 1940s many of the country houses had no screens over the windows to keep out flies from the barnyard and the mosquitoes from the creeks. Many families considered DDT a blessing because it worked better than any other control for nuisance insects. Few had the luxury of refrigerators or iceboxes to preserve fresh food for more than a day or two. The smokehouses of the early 1900s were the main way country people preserved meat for a few weeks or months before insects laid their eggs in the cured sides of pork and infested them with "skippers" (larvae).

The convenience and cleanliness of modern plumbing fades the fact that only two generations ago most farm families in Georgia made do with outdoor toilets, hand-drawn water, and weekly baths in galvanized tubs. Although 73 percent of urban residents in Georgia had water piped indoors in 1940, 95 percent of farm families drew water from a well or spring by hand. Half of urban residences had flush toilets; 97 percent of farmhouses did not.[70] Daily baths and modern plumbing of today probably contribute as much to our improved health compared to the 1930s as any other factor. The family water supply was often a shallow hand-dug well in the midst of the cluster of farm buildings. Frequently, it was adjacent to the porch so as to be handy for the drawing of water. The voluntary, and sometimes involuntary, restrictions on water use of some cities and counties today seem mild compared to the dry wells of the early twentieth century when water had to be hauled for miles by mule and wagon from the nearest spring or large creek.

By today's standards Georgia homesteads of the 1930s and 1940s were not only unhealthy but also unsightly. There were very few green lawns in the first half of the twentieth century, because there were few lawn mowers and little time for tending the yard. Many farmsteads had fruit trees and some had shrubs and flowers, but the extent of "landscaping" of today was unknown and the term was not in the rural vocabulary.

Before the use of lawns in Georgia, yards were swept clean around the neatest homesteads, with gallberry-bush brooms in South Georgia and stiff brooms of other shrubs in other parts of the state. According to Prunty (1965), the clean-swept yards of the early residents of Georgia that persisted into the twentieth century may have originally been for protection of houses from frequent fires set to burn the woods and grazing land of early settlers. In wet weather, old-time yards became muddy; in dry weather they were great playgrounds, quick to soil any kind of clothing. Most people who remember growing up in rural Georgia in the early 1900s would readily agree with Jimmy Carter's (2001) assessment of his boyhood homestead; "I now recall those days with more fondness than they deserve."

The rural homes of today are much more nearly like modern suburban homes than like the farm homes of the past. But even the city homes and yards of the early 1900s were not very sanitary. Garbage collection was not common early in the twentieth century and much of it was burned at home or simply dumped on the backside of the lot. In the bigger cities and in the better neighborhoods it was collected and hauled to an open dump to be burned or scavenged by animals or, in some cases, humans. In the fringes of the city and in the lower class neighborhoods the conditions of the home premises were little different from the farmsteads, except the concentrations of animals were not as great.

At the beginning of the twentieth century the population of Georgia was about 2.2 million, just over one-fourth of that in 2000, and that low population was dispersed over the state much more evenly than today. One of the main causes of pollution is the heavy concentration of population; thus, the low population and its dispersion over the state tended to minimize pollution at the beginning of the century. A popular cliche among environmental scientists "the solution to pollution is dilution" is largely true, and so, early in the twentieth century the sources of pollution were rather diluted. Of course, in the cliche, the dilution refers to the pollutant, but dilution or dispersion of the sources of pollutants also reduces pollution, or stated more correctly in this context, concentration of the sources increases pollution. Of the 2.2 million people in 1900, only 16 percent lived in urban areas of 2500 or more. By 1940, the population had grown to 3.1 million and 34 percent lived in urban areas, so even before mid-century, the concentration of people in urban areas was well under way. In 1970, the state population had grown to 4.6 million and 60 percent lived in urban areas; that is to say 60 percent of the people lived on about 4 percent of the land.

The largest concentration of people, by far, is in the metropolitan Atlanta area, and late in the century sprawl of the suburbs was widely analyzed and condemned. For Atlanta and the surrounding region the balance between benefits and detriments could not be found nor agreed upon. The conflict between benefits and problems is hinted at in a front-page article in the *Atlanta Constitution*.[71] The article related that Georgia had gained 1.3 million people in one decade, a growth rate that was called "staggering." The "staggering" growth was uneven, with 880,000 of the newcomers in the metropolitan Atlanta area. The growth was said to "point up Atlanta's ever elusive struggle to tame its traffic, sprawl and environmental woes," but at the same time was "to the detriment of more rural South Georgia." Would South Georgians consider it a detriment to be free of the "traffic, sprawl and environmental woes" of Atlanta? Not likely—South Georgians more likely consider it detrimental that the concentration of voters and economic power in the Atlanta area grows ever larger.

The crowding of people in cities probably is the basic motivation for a movement in recent decades to set aside land for public use and reserve. Urban people probably feel more need for the set aside of natural areas than do rural residents, who are not so removed from them. In a study of Georgia's environment in the mid-1980s, it was concluded that, "A statewide effort is urgently needed to increase natural area preservation, with the goal of getting 20 percent of the state under some form of protection."[72] The report was pessimistic about the chances, however, because it stated that "Any mention of a need for land-use planning tends to be considered by the highly conservative population as 'communistic.'" It is a measure of the change in attitudes that less than fifteen years later a panel appointed by Governor Roy Barnes recommended acceptance of the set-aside proposed in the 1987 report.[73] There is, of course, some disagreement about what constitutes "green space," but the goal may be halfway met already. If all publicly held lands are added up, according to the latest estimates, there were at the end of the twentieth century about 10 percent of the state held by public agencies (Table 2.2).

One of the motivations for setting aside "public land" is to allow it to remain or return to a wilder state. It is evident to one who grew up in the country in the 1930s and 1940s that an overwhelming portion of the land at the end of the twentieth century is wilder than in the 1930s. The movement of people from the countryside to the cities created economic problems for the rural areas and even greater pollution problems for the cities. But for the rural areas much of the land is better used than in the past and the cities are much cleaner because of regulations and technology. The lead-in quote for this chapter was an admission by Aldo Leopold that the United States had not used its lands decently in the past. There will

Table 2.2 Government owned or managed lands in Georgia.

Source and Agency	Acres	Percent
McCrary and Kundell, 1997.		
Ga. Dept. Natural Resources	1,128,612	3.17
U.S. Army Corps of Engineers	833,917	2.34
Other U.S. Government Agencies	1,380,129	3.70
Total	3,342,658	9.21
Georgia Statistical Abstract, 1998.		
County	188,900	0.50
Federal Land	2,086,500	5.53
Municipal	126,100	0.33
State	782,300	2.07
Water (unspecified)	1,016,300	2.69
Total	4,200,100	11.14

always be controversy about land use because the most-used land affects so many and ideas about its use are so diverse. However, the sweeping changes that have occurred in the use of Georgia's land in the twentieth century must certainly constitute movement toward decency. The great decrease in misuse of the land described in the chapters on soil erosion, water pollution, and wildlife is a cause for pride, if not celebration.

[1] Meine, 1988.
[2] Coulter, 1965.
[3] Coulter, 1965.
[4] Chalker, 1970.
[5] Trimble, 1974.
[6] See chapter 3.
[7] Wetherington, 1994.
[8] Wetherington, 1994.
[9] Wetherington, 1994.
[10] Range, 1954.
[11] US Bureau of the Census, 1920-2000.
[12] Range, 1954.
[13] Range, 1954.
[14] Range, 1954.
[15] Range, 1954.
[16] Range, 1954.
[17] Range, 1954.
[18] Fite, 1984.
[19] Range, 1954.
[20] Winn, 1968.
[21] President's Commission on the Health Needs of the Nation, 1951.
[22] Bledsoe, 1960.
[23] Caldwell, 1930.
[24] Harper, 1923.
[25] US Bureau of the Census, 1990.
[26] Bachtel, 1990.
[27] US Bureau of the Census, 1920-2000.
[28] Citizens Fact-Finding Movement of Georgia, 1946.
[29] Citizens Fact-Finding Movement of Georgia, 1946.
[30] Odum and Turner, 1987.
[31] Range, 1954.
[32] Turner and Ruscher, 1988.
[33] Southeast Region Post-War Planning Committee, 1944.
[34] Georgia Department of Natural Resources, 1996a.
[35] Citizens Fact-Finding Movement of Georgia, 1946.
[36] Prunty and Aiken, 1972.
[37] Johnson, 2000.
[38] Rivers and Loggins, 1989.
[39] Werblow and Cubbage, 1985.
[40] Spillers, 1939a, 1939b, 1939c.
[41] Prunty and Aiken, 1972.
[42] Minor, 1999.
[43] Spillers and Eldredge, 1943; Georgia Forestry Commission, 1946–1999.
[44] Holbrook, 1943.
[45] Clark, 1984.
[46] Cowell, 1998.
[47] King, 1997.
[48] See chapter 6.
[49] See chapter 6.
[50] Kundell and Woolf, 1986.
[51] Georgia Department of Natural Resources, 1999a.
[52] Barrows et al, 1917 and McCallie, 1911.

[53] Georgia Water Use and Conservation Committee, 1955.
[54] Georgia State Soil and Water Conservation Needs Inventory Committee, 1962.
[55] Barrows et al, 1917.
[56] McCallie, 1911.
[57] Kundell and Woolf, 1986.
[58] Kundell and Woolf, 1986, citing a Georgia Department of Natural Resources and Georgia Tech study.
[59] US Department of Agriculture, 1961a.
[60] Seabrook and Soto, 1997; Georgia Department of Natural Resources, 1997.
[61] Georgia Department of Natural Resources, 1999a.
[62] Dahl and Johnson, 1991.
[63] US Bureau of the Census, 1920–2000.
[64] US Department of Agriculture, 1950.
[65] US Department of Agriculture, 1961a.
[66] See chapters 3 and 4.
[67] See discussion of the Sandy Creek watershed in chapter 3.
[68] Kundell and Woolf, 1986; Hefner et al, 1994.
[69] Bartlett, 1993.
[70] Citizens Fact-Finding Movement of Georgia, 1946.
[71] Chapman, 2000.
[72] Odum and Turner, 1987.
[73] Seabrook, 2000b.

CHAPTER THREE

Restoring the Land

"I believe in a permanent agriculture, a soil that shall grow richer rather than poorer from year to year."

—From *The Farmers Creed*, by Frank I. Mann[1]

Erosion: The Great Disaster

The sorriest episode in the environmental history of Georgia was the loss of topsoil from her farmland. The basis for a healthy environment is the land. Most pollution problems impact land and misuse of the land is the first and most basic pollution problem. When productivity of the land is diminished, prosperity and productivity of people are diminished. Most Georgians today do not realize that the productivity of our land is much less now than it could have been. Red knolls in Piedmont fields and forests, from which the topsoil has washed away, produce stunted crops and trees. Deep gullies like Providence Canyon in southwest Georgia ruined the land for farming. These knolls and gullies are testament to the mistreatment of the soil by early settlers and generations that followed. The washing of soil from the uplands impoverished not only the land but also the generations of Georgians that depended on its eroded lands. It was only in the latter half of the twentieth century that the degradation of Georgia's lands has been reversed.

Topsoil is the surface layer of soil, which is the most fertile and best for root growth. When topsoil is deep and fertile it can produce luxurious growth and high yields; when the topsoil is thin, plants struggle for water and nutrients and are stunted and low yielding. The topsoil is the layer most likely to be rich in humus where fungi, bacteria, earthworms, and other soil-borne organisms grow. In the virgin forests of Georgia, the topsoil was deep and rich because of the long period of weathering of minerals, recycling of mineral nutrients, and deposition of organic matter by countless generations of plants and soil creatures. Upon clearing of the forest for cultivation, this layer became impoverished in organic matter, mineral nutrients, and the living things that depend on them.

Worse than that, the removal of the forest subjected the bare soil surface to the impact of rain and erosion. Rain falls on bare soil with surprising force. Although individual raindrops appear gentle enough, the aggregate force of 1 inch of rain on an acre is 113 tons falling from a great height. The effects of gravity do not stop there. The water either soaks into the ground or runs down the sloping surface. Individual raindrops can disperse quite a few soil particles and 1 inch of rain can move a great deal of soil. If rain falls faster than it can soak into the soil, the smaller particles tend to float. When the soil lies on a slope, some of it moves downhill with the water in the process of erosion. Heavy rain on sloping land can produce many small streams at the top of a slope that merge to become torrents at the bottom. Small soil particles are suspended in water running off the surface and flow from the fields with the water. Larger particles too heavy to be suspended are pushed along by the running water like sand and pebbles in a stream.

Plants that cover the soil neutralize the erosive force of water. Well-developed forest with its canopy to intercept raindrops and the heavy layer of litter to protect the soil in winter reduces erosion to nearly zero (Table 3.1). Close-growing plants like pasture grasses and legumes, and even thick growing weeds, absorb the force of raindrops and protect the soil almost as well as forest. From a 1979 survey, the SCS estimated that the average soil loss from cultivated croplands in Georgia was 7.34 tons per acre per year compared to only 0.3 ton for forest, and 0.72 ton for pasture and hay land (Table 3.1). Even the residue from harvested crops protects the soil if it is thick enough. In the conservation practice of minimum or "no-till" cultivation the crop is planted into the residue of last year's crop without plowing the soil and exposing it to erosion. It is the cultivation of clean-tilled crops and exposure of bare soil that constitutes "erosive land use" described by Trimble (1974) as so destructive to sloping soils.

The Uplands Were Wasted

Even though there was substantial clearing and burning of forest land by Indians, there was apparently very little erosion in pre-colonial days.[2] This conclusion is based primarily on the reports of early observers that (1) streams ran clear, (2) bottomlands were well-developed soils without recent sediment overlaying them, and (3) Indians farmed mostly on flood plains of streams where the

Table 3.1 Average rates of soil erosion estimated for Georgia land with different kinds of cover.

Land cover	Year					
	1975	1979	1982	1987	1992	1997
			tons per acre			
Cropland (cultivated)	14.8	7.34	6.3	6.1	5.5	5.9
Forest	-	0.3	-	-	0.0	-
Hay and pasture	-	0.72	0.5	0.4	0.4	0.4

Estimates by the U.S. Department of Agriculture (1977) for 1975, the State Soil and Water Conservation Committee (1981) for 1979, and the U.S. Department of Agriculture (1994, 1999) for 1982-1997.

slopes were only slight. According to Trimble (1974), the American naturalist, William Bartram described Piedmont streams variously as "clear," "crystal," "silvery," and "transparent" during his botanical surveys of the state in the late 1700s. Cleared lands that buffalo were known to have grazed[3] were unlikely to have remained bare of vegetation for a long time, and the grasses and low growing vegetation that buffalo and other animals grazed would have protected the soil almost as well as forest. Certainly the instances of erosion that occurred because of Indian agriculture or forest burning were negligible compared to that caused by wholesale clearing and cultivation that followed settlement by Europeans.

In the process of settling Georgia, especially the Piedmont, land was cleared, farmed to exhaustion, and abandoned as described in chapter 2. The exhaustion included loss of fertility, loss of topsoil, and, for much of the land, deep gullies. The small-scale cultivation of cotton and other crops in the 1700s on the eastern Georgia Piedmont undoubtedly caused localized soil erosion. However, the invention of the cotton gin, the increased industrial demand for cotton, and introduction of slave labor for cotton cultivation gave a strong impetus to the erosive use of so much of Georgia's land. The pattern of erosive land use was prolonged by three major factors, at least two of which followed from the lasting hold of cotton on Georgia agriculture. First, the early settlers believed that the supply of land was inexhaustible, an attitude that lasted beyond the settlement of Georgia. When fertility of cultivated land subsided, they moved on to new, rich soil. Second, the relatively high cost of slave labor compared to cheap land made clearing new ground more economical than care and conservation of land already cleared. Third, following the end of slavery, the tenancy system prolonged the careless use of the land and continued its erosion.

These social and farming systems put a low priority on conservation of land, and the cycle of clearing, cultivation, and abandonment was repeated across the south and much of the rest of the United States from colonial times to the 1930s. The great soil conservation crusader, Dr. Hugh Hammond Bennett, indicted the whole country for land mismanagement (soil erosion was not just a Southern problem) in the 1930s: "The plain truth is that Americans as people have never learned to love the land and regard it as an enduring resource. They have seen it only as a field for exploitation and a source of immediate financial return."[4] In Georgia, the clear-, cultivate-, and abandon system (applied to the tropics as slash and burn agriculture) was at its worst in the lower Piedmont of Georgia until the land was devastated and then it shifted to the upper Piedmont.[5] However, the worst erosion in the upper Piedmont was much later than in the lower part. Trimble concluded that erosive land use in the upper Piedmont in 1920 was equivalent to that in the lower section in the 1860s.

It is difficult to overstate the degree of devastation of the soil. The erosion of croplands began early and lasted 150 years. Muddy streams in early Georgia were both the result and clear evidence of the uplands washing away. The muddying of the streams followed the pioneers westward across the Piedmont of Georgia as is illustrated in the following quote from Sir Charles Lyell in 1849:

> As our canoe was scudding through the clear waters of the Altamaha, Mr. Couper mentioned a fact which shows the effect of herbage, shrubs, and trees in protecting the soil from the wasting action of rain and torrents. Formerly, even during floods, the Altamaha was transparent, or only stained of a darker color by decayed vegetable matter, like some streams in Europe that flow out of peat mosses. So late as 1841, a resident here could distinguish on which of the two branches of the Altamaha, the Oconee or Ocmulgee, a freshet had occurred, for the lands of the upper country (Piedmont), drained by one of these (the Oconee) had already been partially cleared and cultivated, so that that tributary sent down a copius supply of red mud, while the other (the Ocmulgee) remained clear, though swollen. But no sooner had

the Indians been driven out, and the woods of their old hunting ground begun to give way before the ax of the new settler, than the Ocmulgee also became turbid.[6]

It didn't take long for the wasteful use of land to show its effects. In 1856, Frederick Law Olmsted, described the results of a few decades of the careless, exhaustive agriculture in the lower Piedmont of Georgia: "…nearly all the lands have been cut down and appropriated to tillage: a large maximum of which have been worn out, leaving a desolate picture for the traveler to behold. Decaying tenements, red, old hills, stripped of their native growth and virgin soil, and washed into deep gullies, with here and there patches of Bermuda grass and stunted pine shrubs, struggling for subsistence on what was once one of the richest soils in America."[7] The wasteful use of the land started early enough that in some places in the eastern Piedmont it was cleared, reverted to forest, and again reclaimed from "old pine fields" by the mid- to late-1800s.[8] Even in the western part of the Piedmont the devastation came early. Trimble quotes the following observation from Harris County in 1851; "The fields that once brought large and remunerative crops, many of them, are reduced to sedge grass, all scarcified [sic] with gullies." In addition to these two descriptions of the devastation, Trimble gives a clear descriptive account of the extent of erosive land mismanagement in the Piedmont of Georgia in the eighteenth and nineteenth centuries, and even up to the 1930s.

The careless agriculture propagated the notion, if it didn't already exist, that land was an expendable resource to be used and abandoned. Thomas Jefferson is reported to have said: "Where land is cheap and rich and labour dear, the same labour spread in a slighter culture over one-hundred acres, will produce more profit than if concentrated by the highest degree of cultivation on a small portion of the lands."[9] Jefferson's observation probably was accurate for the time. Land was "cheap and rich" relative to labor in the

1700s and 1800s and it probably was profitable in the short run to clear and plant the rich "new ground," use up the natural fertility, and move on.

The urge to move west was not based solely on the "wearing out" of new lands. Many settlers moved westward even in the early days of settlement before the new lands were ruined. This movement was a continuation of the migration that brought the settlers from the Carolinas and Virginia. The parents of George Rockingham Gilmer, one-time governor of Georgia, moved to the Broad River Valley of northeast Georgia from Virginia in the late 1700s. Although Governor Gilmer stayed in Georgia, five of his siblings emigrated to Alabama and a sixth to Mississippi.[10] In fact, Coulter's book is filled with accounts of well-known Georgia families passing through the Broad River Valley on the migration of parents from Virginia and the Carolinas and of offspring to territories farther west.

The tradition of migrant farmers continued long after the initial settlements and was ingrained in Georgians well into the nineteenth century. The poet, Sidney Lanier, knew that tradition because he told a tale in 1869 of a man named Jones from Jones County who had the itch to move on.[11] Two pertinent stanzas follow:

> I knowed a man, which he lived in Jones,
> Which Jones is a county of red hills and stones,
> And he lived pretty much by gittin' of loans,
> And his mules was nuthin' but skin and bones,
> And his hogs was flat as his corn-bread pones,
> And he had 'bout a thousand acres o' land.
>
> So him and Tom they hitched up the mules,
> Pertestin' that folks was mighty big fools,
> That 'ud stay in Georgy ther lifetime out,
> Jest scratchin' a livin' when all of 'em mought

> Git places in Texas whar cotton would sprout
> By the time you could plant it in the land.

The poem is entitled "Thar's More in the Man Than Thar is in the Land" and it is about land mismanagement more than wanderlust, but it captures the philosophy of many landed, and poor, Georgians of the century. Lanier came back to the same theme in his poem entitled "Corn" in 1874 referring to both soil erosion and abandonment about which he was familiar and somewhat passionate.

> Look, thou substantial spirit of content!
> Across this little vale, thy continent,
> To where, beyond the mouldering mill,
> Yon old deserted Georgian hill
> Bares to the sun his piteous aged crest
> And seamy breast,
> By restless-hearted children left to die
> Untended there beneath the heedless sky,
> As barbarous folk expose their old to die.
> Upon that generous-rounding side,
> With gullies scarified
> Where keen Neglect, his lash hath plied,
> Dwelt one I knew of old, who played at toil,
> And gave to coquette Cotton soul and soil.
> Scorning the slow reward of patient grain,
> He sowed his heart with hopes of swifter gain,
> Then sat him down and waited for the rain.

A lesser-known Georgia poet, Henry Rootes Jackson, painted a similarly bleak picture of the mismanagement of Piedmont lands years earlier.[12] In a poem written more out of love than protest, even the title, "The Red Old Hills of Georgia," hints at the devastation. The first stanza makes it clearer.

> The red old hills of Georgia!
> So bald, and bare, and bleak—
> Their memory fills my spirit
> With thoughts I cannot speak.
> They have no robe of verdure,
> Stript naked to the blast;
> And yet, of all the varied earth,
> I love them best at last.

At least one critic[13] thought that "The Red Old Hills of Georgia" achieved as native "lore" almost the status of "The Volga Boat Song" in Russia or "O Sole Mio" in Venice. It is epic irony that the title phrase, known so widely for its symbolism and endearment, was a salute to Georgia's worst environmental disaster. A phrase of shame paraded as pride.

The hills of the Piedmont of Georgia were made red by their careless use. They were once the lighter sandy-beige or brownish color of the topsoil. According to Hartman and Wooten (1935), "There is fairly conclusive proof that all, or nearly all, of the Cecil Soils [The name for the most common soil type in the Georgia Piedmont] in their virgin state had a sandy surface soil. This is shown by many of the remaining areas, which had never been cleared of forest cover. Regardless of how steep the slope, they still retain a sandy covering over the clay loam and clay subsoil." In the area of the Piedmont surveyed by Hartman and Wooten (the old Cotton Belt), they found that the Cecil soil series covered two-thirds of the region. Neither the Cecil soils nor any of the other major soil types had reddish-colored topsoils. Had the topsoil not been allowed to wash away, "The Red Old Hills" would never have come to symbolize the Piedmont, and by extrapolation, Georgia.

Poets were not the only ones concerned about the devastation of Georgia's soils and were unlikely to have been the most knowledgeable about it. However, the above passages show that they knew the reality and the peril of the gross mismanagement. Since they knew, the extent of the problems must have been widely known among the population. Although most may not have known the complex reasons for the massive and persistent misuse of the land, it is now clear that the devotion to "coquette Cotton" that Sydney Lanier spoke about led to the devastating erosion of the uplands, where the first and worst effects occurred. The widespread erosion and abandonment of cultivated lands is testified to by early narrative accounts and later scientific studies. The Piedmont of Georgia is estimated to have lost an average of 7.5 inches of topsoil[14] and, of course, some areas lost all of it. In a land survey in 1934–1936, Spillers and Eldredge (1943) estimated that more than 3.6 million acres in the northern two-thirds of Georgia were in an active state of erosion and 670,000 of those acres were idle or abandoned cropland. A survey of the watershed of Lake Jackson near Covington about the same time revealed that 99 percent of the total area had been affected by erosion including at that time "10.9 percent of idle land usually undergoing serious erosion."[15]

Erosion was worst in the Piedmont, but it was not limited to that area. Land in Southwest Georgia in the vicinity of Providence Canyon was particularly subject to erosion because of sloping land as steep as some of the Piedmont and a poorly consolidated subsurface that was quick to wash away once gullies broke through the surface layer. A publication prepared by the WPA in the 1930s and reissued in 1990 describes the situation as follows. "Half the farms in Stewart County have at least one gully not less than 50 feet deep which has ruined for cultivation at least fifteen acres of land; one such gully, three hundred feet wide and two hundred feet deep, affects more than three thousand acres."[16] This statement undoubtedly refers to Providence Canyon, the largest of the gullies. Half of the farms with 50 foot gullies may be an exaggeration (50 feet is the height of a four-story house.), but some of the gullies were (and still are) spectacular. The area of the most severe erosion was along Turner Creek where there are at least fifteen major canyons eroded into the uplands. The deep canyon-forming erosion was not unique in the Turner Creek watershed, however, but occurred throughout southwestern Georgia and southeastern Alabama.[17] All across the more rolling parts of the Coastal Plain erosion was a problem, although not as bad as in the Providence Canyon area of Southwest Georgia or the Piedmont.

The effects of soil erosion on agriculture are impossible to measure accurately, but they are substantial and long-lasting. Although crop yields have doubled or tripled since the 1930s, yields on eroded spots are today much less than on soils that have not been eroded. One example (Figure 3.1) shows corn yields on different parts of a field where the soil has eroded to different degrees.[18] The thinnest topsoil is on the steepest parts of the slope and the thicker soil is on more level spots or in swales where topsoil has

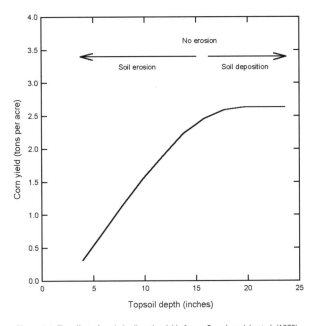

Figure 3.1. The effect of eroded soil on the yield of corn. From Langdale et al. (1979).

washed down from the slopes. Yields are lowest where the soil is thinnest; where half the topsoil is gone, yields are reduced by at least half. Langdale et al (1979) concluded that, although unirrigated corn yields had doubled in the thirty years prior to that time, the effect of erosion on yield had not changed. Loss of 6 inches of topsoil resulted in 40 percent yield reductions in the 1970s, as well as in the 1940s. It cannot be known for sure that those parts of the field with the thinnest topsoil were ever equal to the places with the thickest topsoil today. It is certain, however, that the washing away of topsoil has greatly reduced yields, not only of row crops, but also of the pastures and forests of Georgia.

The Streams Were Spoiled

Although the first and worst effects were on the uplands, bottomlands were devastated as well. The sand, silt, and clay washed off the fields clogged streams and covered the fertile soils of the stream valleys. Deposition of the soil in streambeds raised the level of the stream and the ground water in the valley. The streams that ran swift and had rocky bottoms earlier were raised to the top of the deposited soil and became meandering and muddy. The rising water table caused the soil of many Piedmont valleys to become too wet to plow. The bottomlands that were prized farming lands of the Indians and early settlers became wet and often ruined for cultivation. In this way, removal of soil from the uplands reduced the productivity of both the uplands and the bottomlands.

Although there was little documentation of the damaging effects of siltation of streams early in the history of erosion in Georgia, as soon as scientific study and observation began many examples of the problem surfaced. In a survey of water power in 1880, the Census Bureau report stated about the lower Piedmont of Georgia and South Carolina; "…in many places filling up with detritus…sand and mud…which is washed in from the hill-sides [sic] so that many shoals are being rapidly obliterated, and at many places where within the memory of middle aged men there were shoals or falls of 5 to 10 feet, at present scarcely any shoals can be noticed."[19] In a drainage survey of Georgia, Barrow et al (1917) said this about some Piedmont areas, "Bottom lands now wholly abandoned or cultivated only at great risk formerly were tilled with little expectation of loss by flooding." The following account is given in a 1935 survey of the area that includes the Alcovy River; "Practically all of the bottom land along the Alcovy River has been covered by sand, or the channel has become so clogged with sand that the bottom lands are now swamps, worthless for cropland."[20]

The Alcovy River became a contentious case between the SCS, which planned drainage of the Alcovy Swamps and those who wanted to preserve them as wetlands.[21] Although the swamps of the Alcovy River were described as "a primeval mystery land that was created before the Indians came to Georgia,"[22] Trimble (1970) provides evidence that the swamp was created by upland soil washed into stream bottoms since the Civil War. The SCS proposed channeling of 80 miles of the Alcovy River and two of its tributaries, Flat Creek and Cornish Creek, to drain 4,326 acres of swampland and reclaim it for agricultural purposes. Wharton (1970) objected to the drainage because it would destroy the beauty and the ecological benefits of the swamp, and proposed the Alcovy as the first of Georgia's "scenic rivers." Others objected because of the feared silting of Lake Jackson downstream and the resulting damage to fishing.[23] The ecologists won the battle; no further drainage of the Alcovy swamps occurred. It is, of course, no comfort to those who objected to channeling the Alcovy River that a recent study suggested dredging of silt from Lake Jackson that has washed in mainly from the South and Yellow Rivers.[24]

Whether the silting of the Alcovy River was prehistoric or the result of nineteeth and twentieth century erosion, there were many examples of silting of streams all across the Piedmont of Georgia. A dramatic example is the fate of Scull Shoals on the Oconee River below Athens. The following story is told by Ferguson (1997a). A water-powered cotton mill was built there shortly after 1800. At the time it was the biggest on the Oconee River. It employed 500 laborers and operated 2,000 spindles and looms. "It was the first brick cotton mill in Georgia; it is a candidate as the state's first paper mill. It supported a village with residences, storehouses, and shops." The first mill apparently used the natural shoals with a drop of about four feet, with perhaps a simple diversion of water. A dam was built on top of the shoals in 1860 to a height of 10 feet. By 1887, the drop of water over the dam was already reduced to 4 feet, and therefore accumulated sediment at the base of the shoal was then about 10 feet deep. It appears that a load of sediment from an enormous flood in that same year rendered the mill inoperable. Two years later, the chief of engineers of the US Army stated that the mill's waterpower had failed due to silting of the river below the dam. Today the water of the Oconee River flows across the site of the dam with no perceptible drop. So not only was the millpond filled, but so was the river valley below the dam, meaning that the river is now flowing on a bed of silt at least 14 feet deep. According to Collier (1964), "little remains as evidence that Scull Shoals was once a prosperous community." Indeed, little evidence is visible today that there was even a shoal.

There are numerous reports from across the Piedmont of the same kind of silting that clogged the Oconee and Alcovy Rivers. The Piedmont region was eroded an average of 7.5 inches and in some cases more than 12 inches. Since the lowlands along the streams occupy only about 10 percent of the area of a watershed, sediment accumulated to several yards deep near some streams.[25] Trimble (1974) gives the example of Mauldin's Mill dam built on the Mulberry River in Hall County whose bed was raised 16 feet by erosion sediment. The dam was built in 1865, abandoned because of silting in 1906, and was eventually buried under 4 feet of sediment. The sand and silt in the channel of the Yellow River in

Gwinnett County was so deep in the 1930s that a 15 foot sampling auger would not reach the bottom of the deposit.[26] In the same survey, it was found that Chicken Creek in northern Fulton County and Alaculsa Creek in the southern part of Cherokee County had channel silt that averaged about 8 feet deep.

Many of the streams that were filled with silt were channeled to reduce swamping of the bottomland and to move water more rapidly during rainstorms to prevent flooding. But unless the source of silt was removed or stabilized, channeling was a temporary solution to the problems. In a work plan for the Little Sandy Creek and Trail Creek Watershed in Clarke, Jackson, and Madison Counties,[27] the SCS proposed channeling of Sandy Creek, which had been dredged between 1911 and 1920 "but has become filled with sand." Because the second channeling of Sandy Creek was proposed at a time when attitudes were changing about swamps, the SCS submitted a second revision to the work plan removing the proposed channeling. The case of Little Sandy Creek illustrates the changing nature of stream-bottom conditions in the Piedmont, from the swamping caused by upland erosion, to drainage in the early 1900s, and to later reversion to wetlands. They may yet be drained again, naturally.[28]

The abandonment of millponds like that at Scull Shoals and Mauldin's Mill was repeated many times in Georgia, with the same devastating effect on the areas of Georgia that depended on dams for waterpower. It is not known if the mill in Sidney Lanier's "Corn" was mouldering simply because the "Georgian hill" was deserted by the "restless-hearted children" or because the stream channel had filled with silt to the point of rendering the mill useless. It is likely, however, that many millponds were filled by 1874 when the poem was written. At the beginning of the twentieth century, there were 426 millponds on Georgia streams and about half of them were in the Piedmont.[29] There were eighty-two on the Chattahoochee River and its tributaries alone. Buie (1937), stated that, "A partial survey shows that ninety-five reservoirs in twenty Georgia counties have been completely filled by silting and 110 other reservoirs have been partially filled and are on the verge of being abandoned." Eakin (1939) reported that in the 1930s, "Practically all small reservoirs in the Piedmont and many others of major size that are more than a few decades old were found to be completely filled, except for normal alluvial stream channel through the region of the original pond." Among reservoirs that were completely filled with silt were those formed by Barnett Shoals Dam (45 feet high) on the Oconee River near Athens and Morgan Falls Dam (50 feet high) on the Chattahoochee River near Roswell. Both dams were less than thirty years old. According to Rowalt (1937), another dam near Athens on Sandy Creek was raised three times in twenty years to maintain a head of water for mill power.

Compared to reservoirs in the Piedmont, those in the Mountains and Coastal Plain apparently accumulated little sediment, but for very different reasons. The mountainous streams had only small portions of their watersheds converted to cultivated crops. So, even though many of the relatively level stream bottoms were cleared and cultivated, the hills were not, and streams ran rather clear. In the 1930s survey of Georgia lakes and ponds, sedimentation of the mountain reservoirs, Lakes Rabun, Burton, Nacoochee, and Tallulah, was judged to be negligible based on clarity of the water, forested watersheds, and the observation after emptying the Lake Rabun reservoir that only a few inches of sediment existed.[30]

The filling of reservoirs in the Coastal Plain was much less than in the Piedmont because the slower moving streams carry much less sediment and because Coastal Plain soils have less of the tiny clay particles which stay suspended in water for long periods. South Georgia creeks and rivers also tend to have wider stream valleys where sediment can settle out before entering the streams. The lower sediment load is illustrated by a comparison of Muckafoone

Lake and the Flint River Reservoir next to it. Muckafoone Lake was constructed in 1905 by damming Muckafoone Creek where it runs into the Flint River. In 1920, a dam was constructed across the Flint River and connected with the old dam across Muckafoone Creek. A canal was dug from Muckafoone Lake to the Flint River reservoir so that water flowed from the older to the newer lake. Eakin (1939) describes the difference in appearance of water from the creek and the river where the canal emptied into it as follows. "A sharp boundary line divides the milky to muddy water of the Flint River from the almost crystal-clear water of Muckafoone Creek… The difference in turbidity results from the fact that the 300-square-mile watershed of Muckafoone Creek lies wholly in the Coastal Plain whereas part of the drainage area of the Flint River is in the Piedmont." Hubbard et al (1990) summarized the water quality of Coastal Plain streams they studied late in the twentieth century as follows; "This good quality may reflect land use practices designed to prevent soil erosion, but primarily reflects the Coastal Plain landform shape, which causes sediments eroded from the uplands to be deposited in the riparian zone [area near the stream] before they can enter streamflow."

One problem resulting from the silting of larger streams in the first half of the twentieth century that is not often recognized was the difficulty of crossing. In early Georgia, fording places were popular on rivers and creeks, but where the rivers were too deep to ford ferries were used. Silting of the streams, which made them shallow and created sandbars, endangered the operation of many ferries in the Piedmont. Glenn (1911) described the condition of many streams in a publication about land conditions and erosion in the Southern Appalachian region. Eight ferries in the relatively short stretch of the Chattahoochee River between Gainesville and Atlanta were inoperable or nearly so because of the silting of the river. This passage from Glenn tells the problem in graphic terms:

At Hutchin's, Maynard's, Roger's, Abbot's, and Warsaw ferries the same difficulties from channel filling with sand were found to exist, and at some times of urgent need for ferriage temporary ferryboat channels through the sand had been made with horses and scrapers. These ferries are now being abandoned, and in the places of some of them bridges will be built, though the cost of bridges prohibits their immediate erection by the county at each of the ferries, and much inconvenience to the people on either side of the river is occasioned by this forced abandonment of long-established crossing places.

So the washing of sediment into streams had detrimental commercial and social effects in the past, in addition to the more obvious silting of reservoirs and filling of stream valleys.

The misplaced soil in the stream bottoms cannot, of course, be removed and placed back on the uplands. So the effects of extended erosion of Georgia's farmlands are still evident today and will plague the state far into the future. In Georgia, one can see mature trees growing in deep gullies (Figure 3.2) and old lakes filled with

Figure 3.2. Mature trees in a 30-foot gully at Lexington, Georgia formed early in the 20th century, but now stabilized by kudzu vines.

silt, graphic reminders of a careless use of the land. Even the muddying of streams today is due, at least in part, to silt that was deposited decades earlier. The casual and careless use of land had a long run in Georgia and the rest of the South and, of course, there were not only long-term physical consequences, but psychological and social ones as well. The poor land use that resulted from poverty, ignorance, and carelessness perpetuated poverty, ignorance, and carelessness. So the cycle started by the initial misuse of the land repeated itself for generations in Georgia and across the South. Progress in education, commercial development, and agricultural production not only lagged behind that of the rest of the nation, but increased only slowly until the middle of the twentieth century. Recognition of the social and economic benefits of land conservation was slow to awaken in the South.

Conserving the Soil

But support for land conservation did take root in Georgia. The changes in Georgia agriculture during the last half of the twentieth century have been more sweeping than at any time in history, and a large part of the change is the conservation of the soil. My first notice of the improved conservation of Georgia lands was an article in the 1955 Yearbook of Agriculture, which I read much later, entitled "A New Song on the Muddy Chattahoochee."[31] The title includes an obvious reference to Sydney Lanier's beloved poem "Song of the Chattahoochee" which endeared the river to generations of Georgians. The yearbook article chronicled the decrease of suspended sediment in the Chattahoochee and other Georgia rivers from the 1930s to the 1950s. I knew from my experience as a farm boy in Laurens County and as an agronomist with the University of Georgia College of Agriculture, that the basis for the improvement was mainly the changed landscape. I remembered the rolling fields of cotton and corn from my youth separated only by neighbors' fences and country roads. I remembered too, when gullies were washed in our fields by heavy rain, and afterward we filled them in with mule-drawn scoops. We also repaired terraces with these scoops to minimize erosion from the next heavy rain. When I go back to the neighborhood of my youth, I see that most of the fields are no longer cultivated, but are planted to pasture or pine trees or have reverted to mixed forest.

A second strong impression of the change in landscape came from an aerial photograph of the Athens, Georgia area that I found in the attic of Conner Hall when the University of Georgia hired me in 1968. The photograph was made about 1935, and the most striking features were cultivated land extending from the hilltops to the banks of the Oconee River and very little forest. Thomas Burleigh, a birdwatcher in Athens in the 1920s and 1930s, described the forests of the area as "scattered stretches of mixed pines and hardwoods."[32] Because so many of the cultivated fields of the 1930s and 1940s were replaced with forest and pastures, I knew that erosion was a small fraction of that earlier in the century.

Conservation practices began in earnest before the middle of the twentieth century. It was claimed in the early 1930s that 60 out of every 100 acres of farmland in Georgia were seriously eroded. And although there were efforts at conservation, such as terracing, even before the beginning of the twentieth century, the measures were in most cases half-hearted and less than half-effective. Trimble (1985) relates that "...terracing was almost ubiquitous on the Southern Piedmont by the 1920s," but "[a]ccording to some, the construction and lack of maintenance made these early terraces not a benefit, but instead they were a conservation detriment, doing more harm than good." Although there were individuals who advocated conservation of the soil, there was not a mobilization of public opinion about the dangers of soil erosion and about efforts to control it.

Governmental and Civic Action

In the 1930s, however, both state and federal governments became active in encouraging conservation. In 1934 a federal law was passed establishing the Soil Erosion Service, mainly for the purpose of surveying the problem. In the next year the SCS (now the Natural Resource Conservation Service, NRCS) was established to replace the Soil Erosion Service and charged with instituting programs to conserve the nation's soils. The US Congress passed a law in 1937 authorizing states to establish Soil Conservation Districts for the purpose of soil conservation with the financial aid of the federal government. In 1937 the Georgia General Assembly authorized the establishment of Soil Conservation Districts (later called State Soil and Water Conservation Districts). There were in mid-century, medium-sized green signs beside the highways announcing entrance into these soil conservation districts. The State Soil Conservation Committee was established and authorized under the law to supervise these districts. The importance of the undertaking is emphasized by the fact that Governors E. D. Rivers, Eugene Talmadge, and Ellis Arnold served successively as the first three chairmen of the state committee.[33] In December 1937, ten districts encompassing forty-five counties had been established and "farmers in many other counties had asked for organization."[34] Governor Herman Talmadge proclaimed the first Soil Conservation Week in Georgia in 1949.

As it turned out, the law established to conserve soil did much more. The expanded role for the Soil Conservation Districts was foreseen from the beginning as is evident from a passage in the act:

> It is hereby declared to be the policy of the legislature to provide for the conservation of the soil and soil resources of this state, and for the control and prevention of soil erosion and thereby to preserve natural resources, control floods, prevent impairment of dams and reservoirs, assist in maintaining the navigability of

rivers and harbors, preserve wildlife, protect the tax base, protect public lands, and protect and promote the health, safety and general welfare of the people of the State.[35]

The early activities of the Soil Conservation Districts did not turn out to be as broad as the last few phrases suggest, but the districts were involved in a wide range of activities. Working with the SCS they coordinated such diverse activities as terracing the land, building farm ponds, improving pastures, draining swamps, managing forests and wildlife, and promoting soil conservation education and incentive programs throughout the state.

The work of the SCS and the state Soil and Water Conservation Districts is a good example of the cooperation of the federal and state governments to solve a national problem. The SCS provided assistance to individual landowners and the programs, both large and small, were coordinated through the state committee. The SCS agents located in nearly every county in Georgia assisted individual farmers in developing farm conservation plans and gave advice on the many practices that reduced soil erosion. By 1950, there were 61,926 farms totaling over 12 million acres—more than half the farmland in the state—actively carrying out detailed plans prepared by the SCS. Over 6 million acres had been treated for erosion and 850,000 acres of pasture had been improved.[36]

The Soil Conservation Districts along with the help of the SCS also organized projects based on whole watersheds and implemented several practices on cooperating farms in the watersheds that contributed to soil and water management and modernization of farming. The first in Georgia was on the Sandy Creek watershed near Athens.[37] In 1934, the Soil Erosion Service (replaced in 1935 with the SCS) entered into five-year agreements with farm owners in a demonstration project to carry out certain land-use practices. In turn the government built check dams for flood and sediment control, established contour strips and constructed terraces, planted trees, seeded eroded areas and performed other necessary operations. They even built fences. Road banks in the watershed were sloped and planted to sericea lespedeza, a legume that grows well on poor, eroded soils.

Large watershed projects were started as a result of the federal Flood Control Act of 1944. The first one organized in Georgia was the Coosa River Watershed and it included twenty counties and over 1.5 million acres. Cooperating farms in the watershed were helped to start conservation practices such as contour tillage, strip cropping, terracing, and farm ponds. All of these were practices designed to hold the soil in place and keep it from washing into streams. Later, many small watershed projects were established in Georgia to accomplish the same purpose of the larger ones.

Demonstration projects emphasizing conservation farming were conducted on farms and groups of farms to show that diversification and conservation not only saved the soil, but actually increased income and improved living standards. From 1938 to 1941, the SCS purchased three large tracts of "submarginal" land in Georgia to "be used for demonstrations of proper use of eroded land and to help relieve social and economic maladjustments in the areas."[38] One tract was in Whitfield County (12,000 acres), one in Greene County (25,000 acres), and the third in Jasper, Jones, and Putnam Counties (68,000 acres). In addition, the Georgia Association of Conservation District Supervisors, at their second annual meeting in 1944, gave their support to a plan for farm improvement in Georgia called Georgia Better Farms, Inc. This plan was the idea and personal project of textile manufacturer Cason J. Callaway of Lagrange. It provided for the purchase of 100 farms in Georgia to demonstrate modern farming methods and placed a strong emphasis on soil conservation.[39]

Another way that the SCS and the districts spread education about conservation was through field days. One of the most impressive was held by the Soil and Water Conservation Districts in

Barrow County on 12 May 1948.[40] The field day was sponsored jointly by the Oconee River District, the *Atlanta Journal,* and the civic clubs of Winder. More than 60,000 people attended. On that day, an army of 800 workers converted the rundown farm of Marion H. Carlisle and Ernest C. Blakely into a conservation "miracle." Governor M. E. Thompson and Dr. H. H. Bennett were featured speakers. Five years later district and agricultural leaders reassembled at the farm to assess the results. Farm income had doubled, but income per farm worker had increased four-fold. Before the transformation cotton and milk were the main products, worth $7,000. Afterward no cotton was produced, but the value of the milk products alone was $13,000. There were also large field days at the Georgia Baptist Children's Home near Baxley and at Reinhardt College Farm in Waleska.

The soil conservation crusade was at its most active from the 1940s to the 1960s when it enlisted many individuals and organizations. During the first "Soil Conservation Week" declared by Governor Herman Talmadge in 1949, according to a newspaper account, "734 newspapers reported the event, 383 clergymen delivered sermons, 93 civic clubs sponsored programs, 87 conservation demonstrations were staged and 245 meetings featuring conservation were held."[41] Susan Myrick, associate editor of the *Macon Telegraph,* and Channing Cope, columnist for the *Atlanta Constitution,* were given awards as Conservation Man-of-the-Year in 1948 by the State Soil and Water Conservation Districts. Myrick received the award again in 1957. The role of churches in the crusade emphasized the moral imperative of soil stewardship. The Georgia Association of Soil and Water Conservation District Supervisors and the individual districts had ministers as chaplains (and still do) who stressed personal stewardship obligations. During the week of 22–29 May 1960, there were 1,072 sermons to 130,982 churchgoers, and in 1967 "more than three-fourths of a million Georgians were involved in these observances, and over 1,800 stewardship sermons were preached."[42] In rural Georgia, soil conservation was serious business.

Success in soil conservation came rather rapidly (see Figure 4.1) and over the years changes in policy and law broadened the activities of the state and federal conservation agencies beyond those of the 1930s when soil erosion was the main menace to farmers. At first the broadening amounted to shifting the emphasis from soil conservation to land management and increased production.[43] Eventually it went even further. The extent of broadening is illustrated by two examples from the 1960s and 1970s. In a 1966 Watershed Work Plan for the Little Sandy Creek and Trail Creek watershed near Athens, recreation was the costliest item and was calculated to give the greatest benefits.[44] The costliest structure was a multipurpose lake with a recreation area. Construction of the recreation area was estimated to cost nearly 1 1/2 times as much as the lake. The estimated annual benefits for recreation were 72 percent of the total benefits for all of the structural improvements in the project, including the reservoir, five flood-retarding dams, and 14 miles of channel improvement. The lack of emphasis on erosion was made clear by a statement in the plan; "There are no agricultural water management problems that would merit consideration of this project."

A second example with a similar focus is the results given by Sisk (1975) for a Resource Conservation and Development project in Gwinnett County authorized under the Food and Agriculture Act of 1962. In addition to the usual conservation plans and construction of ponds, "To provide hunting, two shooting preserves have been established, thirty dove fields, two duck ponds and one rabbit ranch." "Forty-five community recreation areas were developed including Little League ball fields, midget football, soccer fields, facilities for baseball, track and tennis, picnic areas and swimming pools." "Set forth in the RC&D work plan was the objective of securing new industries to provide 10,000 new jobs. This has been exceeded." Held and Clawson (1965) characterized efforts to broaden the scope of soil conservation in terms of both trends and politics. They said the changes "have been more a matter of the established soil conservation programs catching on to popular bandwagons than of any fundamental broadening of soil conservation as such." They further stated that "a program with something for every area, and if possible for every farmer, was politically if not technically necessary." Whatever the cause, the inclusion of so many activities not closely related to saving the soil indicates a lack of urgency barely thirty years later about the problem for which the agencies were created. The success of soil conservation in three decades lessened the demand for remedy and changed the governmental focus toward other issues.

What are the Results?

In spite of the considerable evidence that soil erosion is a small fraction of that early in the twentieth century, the great success in stopping erosion after 150 years of soil mismanagement has gone largely unheralded. Although there are many indirect examples of the improvement, few quantitative studies have measured the effects of this long-standing conservation program.[45] Held and Clawson attempted a national assessment of the progress in soil conservation in 1965. They expressed the following frustration at not being able to adequately measure the progress; "There are simply no clear-cut, well-designed data, especially on a historical scale, which are adequate to answer the critical questions about soil conservation achievement over the past 30 years or so." They also added, "In view of the sums of money annually spent for soil conservation and the relatively long period during which such a program has been carried on, it is high time more is known about what is being done." At the end of the twentieth century a better, though still circumstantial, case can be made for success of the soil conservation movement in Georgia. The circumstances that led to such devastation of our land have greatly improved and it can be concluded by inference that so has the conservation of the soil. Although Bennett (1955) announced that "...the Soil Conservation Service has compiled a great store of factual information about the nation's land resources, the problem of erosion, and the methods of soil and water conservation," there is no great store of knowledge about the accomplishments in soil conservation. But the evidence is plentiful.

Covering the Bare Places

The most visible evidence of decreased erosion is a decrease in the amount of soil exposed. Because erosion from an acre of cultivated cropland is more than ten times that from pasture and twenty times more than from forest (Table 3.1), erosion in the state is nearly proportional to the amount of land in cultivated crops. It is easy, then, to conclude from Figure 2.1 that erosion decreased by at least three-fold in the twentieth century because the row crop acreage decreased by that amount. But the decrease in erosion is greater than three-fold because late in the twentieth century row-crop land was less erodible than early in the century. Erosion on cropland has greatly decreased (Table 3.1). Better conservation practices, including less tillage of cultivated acres, were used in the 1990s, and the extent of eroding, abandoned land was much less. Trimble (1974) calculated that, "If, for example, row crops were considered to have an erosive value of 1.00 about 1920–1930, the present relative value would be reduced to approximately 0.45 because of terracing, contouring, cover crops, and other conservation practices."

Although erodibility of cropland was reduced by more than half from the 1920s to the 1970s, the improvement did not stop there. In 1975, the SCS estimated that the average yearly erosion on cropland in Georgia was 14.8 tons per acre.[46] Actually the 14.8 tons per acre is an average of estimates from three regions (14.9, 13.4, and 16.1 tons per acre) that cover the state and parts of adjoining states. In 1979 the SCS calculated that an average of 7.34 tons of soil eroded from cultivated cropland in the state. By 1987, it had dropped to 6.1 tons per acre, and in 1997 to 5.9 tons per acre, a drop of 60 percent in 22 years (Table 3.1). The change in erosion rate from 1975 to 1979 is unrealistically large. In reality, all of the numbers in Table 3.1 are calculated estimates that probably overestimate the soil removed from fields.[47] They do reflect, however, the progressive conviction among soil conservationists that cultivated land is not as erodible as it was early in the twentieth century.

One reason that cultivated cropland in Georgia is less erodible today than early in the twentieth century is that the most erodible land was taken out of cultivation. In the severely eroded Providence Canyon region of Southwest Georgia, cropland area decreased from 60 percent of the landscape late in the nineteenth-century to nearly zero in the 1980s.[48] Highly erodible land continued to be taken out of cultivation even late in the twentieth century and at the end made up only about 10 percent of the cultivated land (Table 3.2). From 1982 to 1997 the acreage of total cropland (including orchards, etc.) decreased by 29 percent, but on highly erodible land it decreased by more than 50 percent. The shifting of row crops to less erodible land was due in part to their movement to South Georgia.[49] In 1935, 40 percent of the harvested cropland was north of the Fall Line, but in 1997 only 14 percent was. Figure 2.3 shows that even though the acreage of row crops has decreased in South Georgia (an average of 34 percent in the eight counties), the decrease was small compared to that in Piedmont counties (87 percent). The fields of South Georgia are less steep and water infiltrates into the sandy soils more readily, both of which reduce runoff of water and therefore erosion of the soil. In the soil and water conservation needs projected for 1975, erosion was considered the dominant problem for 93 percent of the cropland in the Piedmont counties represented in Figure 2.3, compared to only 65 percent in the Coastal Plain counties.[50] The shifting of row-crop agriculture to South Georgia was the result of a combination of factors, includ-

Table 3.2 Acres of cropland in highly erodible and less erodible classes from 1982 to 1997. From the U.S. Department of Agriculture, 1998.

	Highly erodible	Less erodible	Total
	------------------------ 1,000 acres ----------------------		
1982	1,000	5,569	6,569
1987	875	5,034	5,909
1992	664	4,509	5,173
1997	483	4,178	4,661
Decrease 1982-97(%)	517 (51)	1,391 (25)	1,908 (29)

ing economics, land suitability for cultivation, and availability of water for irrigation.

A second reason for the progressively lower erosion rates is improvements in soil conservation, long promoted by the SCS and Soil and Water Conservation Districts. This promotion included improved tillage. Early in the century, tillage was done without much regard for erosion of the soil. The soil surface was cleared of any crop and weed residues before planting. It was common for crop residue from the previous season to be raked and burned in preparation for spring plowing. Crop residue was considered trash that obstructed the preparation of a good seedbed. Terraces to divert rain water slowly around the slopes were either not used, or as indicated above, they were poorly constructed and maintained. Rows of plants frequently ran up and down the hills, rather than on the contour, so rainwater and soil could more easily flow downhill rather than remain in place. Tillage of a single crop was intensive before the 1950s, with the mule and plow passing up and down the rows as many as four or five times after the crop was planted. This was in addition to two passages of the hoe-hands, thinning the cotton plants and removing weeds within the row.

The idea began to grow about the middle of the twentieth century that the old cultivation involved too much tillage. In addition to promoting better terracing, protection of natural drainage ways in the fields with grass sod, and plowing on the contour, conservationists began to emphasize "conservation tillage." In conservation tillage the soil is tilled less and crop residues are left on the surface to absorb the erosive effects of rainfall. Research has shown that conservation tillage reduces erosion by 50 to 95 percent compared to conventional clean tillage[51] depending mainly on the amount of plant residue left on the surface. Conservation tillage is defined as "any tillage method that leaves at least 30 percent of the soil covered with crop residue immediately after planting."[52] Today a large percentage of Georgia's crops are grown with little or no tillage at all, before or after planting. The Georgia Nonpoint Source Management Plan (1989) states that, "—the amount of cropland and pastures in no-till cultivation has increased from 580,000 acres (11% of the total crop and pasture land) to 1,300,000 acres (31% of total crop and pasture) in 1988 according to information provided by [the Georgia Soil and Water Conservation Committee]." In the no-till system, the seeds are planted in the residue of the previous crop and weeds are controlled with herbicides. In such a system soil erosion is reduced to levels close to that on pasture or forest. Winter pastures of rye, wheat, or oats formerly planted on thoroughly plowed land is now more often planted with a "no-till drill" directly into existing sod. Such a practice practically eliminates erosion that might occur before the winter grasses are established.

Although the greatest decrease in soil erosion has resulted from changes in agriculture, other practices also have contributed to saving the soil. Paving of roads and sodding of highway slopes have greatly reduced the amount of mud that washes off the roads and into streams. In 1941, Georgia had 104,513 miles of roads and 70 percent were unpaved.[53] Rural areas had the fewest paved roads. In 1930 only 6 percent of Georgia's farms were on hard-surfaced roads; and as late as 1950, more than 70 percent of farms were still on dirt or unimproved roads.[54] Chapman et al (1950) estimated that at mid-century Georgia's roads, including rights-of-way, occupied about 700,000 acres of land, and much of it was nearly completely and continuously unprotected from the erosive effects of rainfall. This area is about equal in size to that of the peanut crop in Georgia in the 1950s. Ditches along both sides of these roads formed ready channels for washing of sediment into streams. As the ditches filled with sediment, county road crews plowed them out bringing the material back onto the roads, only to be washed off again in the next series of rains. The paving of 30 percent of roads by the early 1940s afforded some protection, but because the backslopes and rights-of-way were often bare of vegetation, they were still subject to considerable erosion.

Soil loss from road slopes early in the century occurred at a rate equal to or greater than from agricultural lands. Chapman et al (1950) concluded that "If all the road rights-of-way were completely stabilized and adequately vegetated outlets for road water provided, a major contribution to a reduction in erosion damage would result." SCS engineers calculated the soil loss rate to be about 500 tons per mile of road annually in Cobb County.[55] In the Coosa River region of northwest Georgia, it was found that on unprotected roadbanks "soil losses averaged over 300 tons per acre—an annual rate some 15 times greater than the annual average rate from sloping croplands in cotton on terraced lands."[56] These scientists concluded that a contributing cause of the high erosion rates was the soil-loosening action of icicles formed during the freezing of wet soil surfaces, especially on north-facing banks. As late as 1973, Governor Jimmy Carter said about erosion and sedimentation, "The Transportation Department of Georgia is one of the most permanent and serious abusers of the quality of streams in Georgia." Six years later, the SCS estimated that 5,651,000 tons of soil were lost per year from 439,000 acres of roads and railroads, an average rate of 12.9 tons per acre per year.[57] By that time many of the roads and roadsides were already protected.

Unpaved roads at the end of the twentieth century were a small fraction of the total, and erosion from roadways was trivial compared to that at mid-century. Nearly complete vegetative cover prevents erosion and protects roadsides today. Not only are the modern roadways not subject to erosion, but erosion during and immediately after construction is minimized by straw or biodegradable fabric laid on the bare areas to protect the soil until vegetative cover is established. Any sediment not held in place by the straw and fabric is likely to be caught before it enters a stream by silt fences constructed around the construction sites. These practices for prevention of soil erosion and sedimentation at highway construction sites did not exist in the first half of the twentieth century.

Along with the reduction of erosion came a beautification of Georgia's roads. In contrast to the "blots on an otherwise fair and attractive landscape,"[58] the roadsides are now lined with grassy slopes and trees that have transformed the gullies and bare roadbanks of "The Red Old Hills of Georgia." In 1931, Mrs. Richard P. Brooks gave a "Highway Beautification Address" to the "Patriotic Citizens of Georgia" exhorting them to a more beautiful landscape. This address, sponsored by the Atlanta Women's Club, outlined three main features for highway beautification; 1) The preservation of trees and forest, especially trees along the highways, 2) Removal of all unsightliness and rubbish from the roadsides, and 3) Planting of trees, shrubs, and flowers. To a large degree, the objectives of that address have been met. Most of the highways of Georgia have pleasant vistas of forests and fields. Some stretches even have wildflowers planted for color. Many of the larger urban and rural intersections now have shrubs and trees planted in pleasing patterns. The Adopt A Highway program to involve citizens in removal of trash and other efforts by the Georgia Department of Transportation have made the roadsides less cluttered. Although it is difficult to convince those who did not know Georgia's roads in the first half of the twentieth century, the dreams and efforts of Mrs. Brooks's and like-minded citizens have made the roadsides their most beautiful ever. Her hopeful rewrite in 1931 of the first stanza of "that dear little poem" of Henry Rootes Jackson is now much more descriptive than the original.

> The old red hills of Georgia,
> Covered over with waving green,
> With highway, hedge and valley,
> The loveliest ever seen.
> Clad in her robe of verdure,
> Jeweled with nature's flowers,
> Go where you will—
> There never can be found,
> A lovelier State than ours.

The view from the roadsides is only a signal of the transformation of the countryside. Some of the land diverted from row-crop production has gone into city and residential sub-divisions, industrial areas, and highways, but most of it has become forest, and nearly all of it has shifted to uses that cause much less soil erosion than row crops and neglected roadways. Kundell (1989) described it well, "In the increasingly reforested Piedmont, the great cotton vistas and the eroded red hills that had characterized Georgia's landscape for more than a century became just a memory." One doesn't have to rely on the facts and figures of foresters and agriculturists to see that the land has reverted from erosive use to more natural vegetation. It is more convincing to drive Interstate Highway 20 from Atlanta to Augusta through the heart of the old Piedmont cotton belt and look at the landscape. An even more dramatic illustration of the greening of Georgia is to compare the video that precedes the viewing of the famous painting of the Civil War Battle of Atlanta with the painting itself. The painting at the Cyclorama shows the hills around the city "so bald and bare and bleak," and the video of the reenactment of the battles leading up to the climax in Atlanta shows no red hills, but trees and green grass. The reenactment was in the 1990s, 130 years after the battle.

The "Red Old Hills of Georgia" have gained a "robe of verdure." It is obvious that only a tiny percentage of the soil is bare. If the view from Interstate 20 is considered to be unrepresentative of the landscape, a drive down country roads shows that the hills are covered with woods and pastures and protection of the soil is nearly complete. Such a reversion of the cultivated landscape led Trimble (1974) to observe in the early 1970s that "Declining agriculture and a strong soil conservation program have generally reduced erosion to a magnitude closer to aboriginal conditions than to that typical of the most intensive erosive land use." The "most intensive erosive land use" Trimble refers to was cultivation of clean tilled crops of cotton and corn in the 1800s and early 1900s. Even as early as 1962, Dr. O. C Aderhold, president of the University of Georgia, observed that "Once barren hillsides are now sodded or in trees. Contoured farming is practiced. Farm ponds for conserving water and providing recreation dot the landscape." The progress has not slowed since the 1960s, and at the end of the twentieth century achievements in soil conservation would make the father of soil conservation in this country, Dr. Hugh Hammond Bennett, proud.

Reducing the Silting of Streams and Lakes

One of the ways of documenting the decrease in erosion is to examine the deposition of silt in stream bottoms and its removal. Most of the sediment washed from the uplands of the Piedmont remains in the stream valleys. Considering the example of Scull Shoals and the other places in the Piedmont that are silted to similar depths, the amount of original topsoil now in stream flood plains is nearly beyond comprehension. It has been estimated that only 4 percent of the sediment eroded from the Piedmont has been transported past the fall line.[59] When the stream valleys were filled in the nineteenth and early twentieth centuries, silt was deposited across the whole width of the valley up to the level of sedimentation. Running water, because of its turbulent energy, has the capacity to carry a certain amount of sediment. When the soil washed off the uplands was more than the streams could carry, the silt accumulated in the streambeds and raised them to higher levels. When erosion was checked in the middle of the twentieth century, the streams were no longer saturated with sediment and they began to pick up sediment from the stream bottoms and to cut new channels into the elevated streambed. The new channels formed in the accumulated sediment are narrow and rectangular. That is, they have rather vertical sides even when they are several feet deep. This is probably a result of the ease with which water cuts through the silt, the fact that the channels are quite recent, and they have been stabilized by vegetation, including trees. It is common, however, to see the streams undercutting the banks and collapsing them along with trees growing on the banks (Figure 3.3).

The decrease in the sediment suspended in streams (Figure 4.1) indicates that much less of it is deposited than years ago. Little direct study of sediment in stream valleys has been done, but the few examples also point to decreased deposits. A study of the Soap Creek bottoms in Lincoln County showed that sedimentation of the floodplain averaged 0.44 inches per year from 1860 to 1923, but less than 0.20 inches per year from 1923 to 1996.[60] Erosion had not been reduced very much for the early portion of the 1923—1996 period, and current rates are likely to be lower. Low sediment deposition in the last half of the twentieth century is also indicated by the fact that "Within the last fifty years, Soap Creek has responded to reduced sediment yields associated with natural revegetation and soil conservation practices by incising its channel back down into the historical alluvium." Another recent study showed that sedimentation in the Ichawaynochaway Creek bottoms

(A)

Figure 3.3. Sloughing of the banks (A) and undermining of mature trees (B) by Brooks Creek in Oglethorpe County.

in Southwest Georgia occurred at a rate of 6.33 tons per acre per year over the last 125 years. In the last thirty years the rate was only 0.37 tons per acre per year, a reduction of 94 percent.[61] At one location in Providence Canyon, it was determined that sedimentation had decreased from 8.25 inches per year for the period 1942–1948 to 0.55 inches per year from 1948 to 1983.[62] So, in spite of occasional statements such as a recent one in the *Atlanta Constitution* (1999) that "Streams and rivers in the region are filling up with so much dirt that they often can't absorb sudden rainfalls, leading to flooded homes and streets," the opposite is true. In the majority of streams sediment is washing out of their channels rather than into them.

An example of the washing of sediment out of Piedmont stream channels is drawn in Figure 3.4.[63] This stream is the Mulberry River in Hall County and is part of the Upper Oconee River Basin. The measurements of the stream valley width and depth were made in 1938 in preparation for building of the Georgia Highway 211 bridge. When Burke measured the channel at the same location in

(B)

The height of the bank in (A), about 6 feet, is roughly the depth of sediment deposited in this stream in the past.

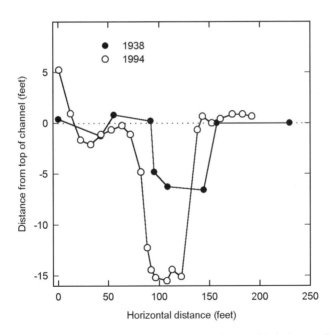

Figure 3.4. Profiles of the Mulberry River channel in Hall County. The tops of the banks were adjusted to be nearly equal to show differences in the depth of the channel. Redrawn with permission from Burke (1996).

1994, its depth had doubled and its capacity had increased by 89 percent.

The cutting of stream channels is part of a long-term pattern of movement of the sediment accumulated in stream valleys. The cutting of the new channels begins in the upper reaches of watersheds in the smallest tributaries. The sediment eroded from the small stream channels moves downstream and the accumulated load prevents the larger streams from cutting into the accumulated sediment as much as in the upper streams. When the smallest tributaries cease to contribute significant amounts of sediment to the larger ones, then the larger streams begin to form deeper channels in the sediment accumulated in their valleys. In this progressive way the sediment in stream channels of Piedmont watersheds is presently migrating downstream.[64]

Even the streams that have cut deep new channels into the accumulated sediment are still carrying sediment downstream. As the streams meander they cut into the steep banks causing erosion and sloughing of the sides (see Figure 3.3). During floods when the channel fills or overflows, the channel sides are scoured carrying large amounts of sediment downstream. So, even though the amount of sediment washing off the uplands is a small fraction of that in the early 1900s, there are still large amounts of sediment migrating down the large watersheds. This migration of sediment down the watershed keeps the largest streams from cutting through the accumulated sediment and returning to their original streambed. As Ferguson (1997b) puts it for Scull Shoals on the Oconee River; "The incision process here is lagging behind that of typical headwaters, because the secondary sediment from the eroding headwaters continues to pour down upon the main stream, slowing net sediment removal." The control of erosion has made the streams run relatively clean once again, but there are long-term consequences for past mismanagement. The sediment slowly moving down the watersheds will remain a problem for centuries as it continually moves downstream.

The ability of a stream to pick up sediment from the streambed and bottomland is illustrated by examples from the Savannah and Etowah Rivers (Figure 3.5 and 3.6). In the years after construction of Clarks Hill Dam (Thurmond Dam) above Augusta in 1952, sediment load in the Savannah River below the dam dropped from between 400,000 and 600,000 tons per year to less than 200,000 tons per year. However, the sediment loads at Clyo, 170 miles downstream, decreased only slightly during the next ten years (Figure 3.5).[65] Since little sediment washes into the Savannah River from the Coastal Plain, the conclusion by Meade was that much of the suspended sediment at Clyo was from that deposited

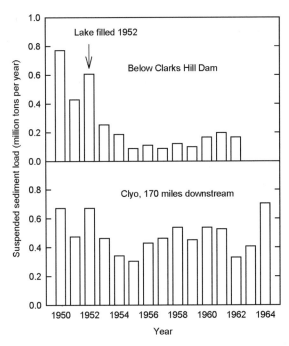

Figure 3.5. Differences in the amount of sediment carried by the Savannah River below Clarks Hill Dam and at Clyo 170 miles down stream for ten years after building of the dam. From Meade (1982).

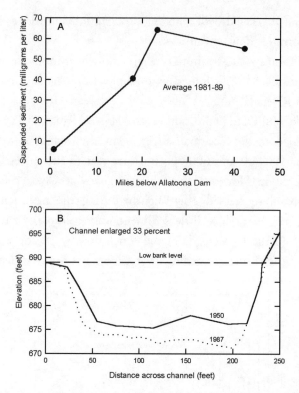

Figure 3.6. Increases in sediment concentration on the Etowah River downstream of Allatoona Lake (A), and the change in the profile of the river channel 3.8 miles downstream from the dam (B). Sediment data from the US Geological Survey (years 1981–1989) and river channel measurements from the US Army Corps of Engineers (2000).

earlier in the channel and bottomlands. In a similar way, trapping of sediment by Lake Allatoona causes the Etowah River to run over the dam relatively clear of sediment (6 milligrams per liter), but at Kingston, about 25 miles downstream, sediment is increased to 64 milligrams per liter (Figure 3.6A). Most of this extra sediment presumably comes from the river channel because no major tributaries enter the river in this stretch. This washing out of the sediment caused about 20 percent enlargement of the channel in the 10-mile stretch below Lake Allatoona between construction of the dam (1949) and the mid 1980s. The enlargement of the channel shown in Figure 3.6B is actually 33 percent, but in other places it is much less. Bagby (1969) understood the potential for resuspension of stream-bank silt in the Yellow River in expressing his fear of the silting of Lake Jackson; "sediment already in the stream is stirred up anew with each heavy rain, slowly washing its way downstream into Lake Jackson for years to come."

As mentioned for the Alcovy River and Sandy Creek above, the SCS proposed enlarging many stream channels all over Georgia to reduce flooding and convert wet bottomlands for cultivation or timber production. Many of the bottomlands were drained by artificial channeling, but many of those that were channeled had their channels refilled, as was the case for Sandy Creek. However, the natural cutting of new channels in the suspended sediment is likely to have much the same effect on wetlands as artificial channels. The lowering of the stream bed lowers the underground water level in the smaller stream valleys, just as the silting of the streams raised it decades ago. As the process of sediment migration described by Ferguson continues and the stream levels are lowered, wetlands along the streams are likely to become drier. In fact, Burke (1996) concluded that the smaller stream bottoms in the upper Oconee River basin are already becoming drier. So the channelization proposed by the SCS for so many streams in the Piedmont may be accomplished without the engineering proposed for those projects long ago. If the Alcovy River swamps were created by manmade sedimentation as Trimble (1970) contended, they may be at least partially drained by the Alcovy River itself, given enough time.

Even though there are still tremendous amounts of eroded sediment in the stream valleys, and even though some of this sediment is migrating downstream toward lakes and ponds, the amount being deposited in these reservoirs is much less than early in the twentieth century. Sedimentation is a subject of occasional speculation from those who worry about contamination of Georgia's lakes, but the authorities seem not to have been very concerned. Although the state has a large number of lakes and ponds, surveys of sedimentation have been patchwork rather than systematic. Few were made in the last half of the twentieth century, and almost none have been published. In surveys of sedimentation, elevation of the lake bottom is measured along ranges from shore to opposite shore across the lake and a calculation of the area of the profile of the range up to the some level is made (usually the high- or normal- water level). If the area of the profile decreases in later surveys, it means that some filling with sediment has occurred along the lake bottom. If enough of these ranges are run to represent the different widths and depths of the lake, an estimate of sedimentation can be made.

There were surveys of a small number of reservoirs during the middle of the twentieth century.[66] Most of these surveys were not repeated so that change in sedimentation can not be traced, but for the few smaller lakes that had at least three surveys, there was a decrease in sedimentation with time (Table 3.3). The exception was flood control dam number 14 on the North Fork of the Broad River where sedimentation increased from the 1950s to the 1960s. A total of 1,349 of these flood control dams or "flood retarding structures" were built in watershed projects across Georgia to reduce flooding and also to trap silt that would otherwise move continually downstream.[67] From 1924 to 1937 the Newnan municipal reservoir on Bolton Mill Creek filled with sediment at a rate of 0.50 percent of the lake volume per year, but in the next nine years the rate decreased by two-thirds to 0.15 percent per year (Table 3.3). According to Brown (1948), this was due to reforestation and gully control in the 890-acre watershed above the reservoir. The large lakes in North Georgia that had three or more surveys showed almost no change with time (Table 3.3). As mentioned earlier the mountain lakes, including Lake Nottely and Blue Ridge Lake, never accumulated much sediment. Lake Jackson in the lower Piedmont was reported to be filling with silt at the same rate as early in the twentieth century, but there is a question about the accuracy of the surveys (see below). None of the reservoirs in the most recent surveys are filling at rates close to most lakes early in the century when Barnett Shoals Lake on the Oconee River and Morgan Falls Lake on the Chattahoochee River filled with silt in less than thirty years. There are no recent surveys for the reservoirs in Table 3.3 reported by Dendy and Champion (1978) and it must be assumed that sedimentation is not considered to be a serious threat in those lakes.

Table 3.3 Loss of water storage because of sedimentation of reservoirs.

Reservoir and location	Drainage area sq. mi.	Period	Loss of storage % / year	Source of data
Lake Nottely	207	1942-55	0.07	Dendy and Champion, 1978
Blairsville		1955-65	0.03	"
Blue Ridge Lake	227	1944-54	0.07	"
Blue Ridge		1954-68	0.003	"
NF Broad #2[a]	0.94	1956-59	0.26	"
Denman's Creek		1959-70	0.16	"
NF Broad #6	3.50	1956-59	0.45	"
Bear Creek		1959-70	0.23	"
NF Broad #11	3.67	1956-59	0.64	"
Tom's Creek		1959-70	0.16	"
NF Broad #14	1.19	1954-62	0.25	"
Tom's Creek		1962-69	0.49	"
Lake Allatoona	1,100	1949-83	0.134	U.S. Army Corps of
Acworth		1983-98	-0.083	Engineers, 2000
City reservoir	1.39	1924-37	0.50	Dendy and Champion, 1978
Newnan		1937-45	0.15	"
Lake Jackson	1,414	1910-35	0.51	"
		1935-90	0.51	Ga. Dept. Nat. Res., 1990b

[a] NF Broad; a series of flood retarding dams on tributaries of the North Fork of the Broad River in Northeast Georgia.

Surveys by the US Army Corps of Engineers on large lakes that they manage also show that recent sedimentation is low. Clarks Hill Lake, which was built in 1952, was surveyed for sedimentation in 1999.[68] The survey showed that "less than 0.3 percent of the total conservation storage" had been lost in that forty-seven year period. Large lakes naturally sediment more slowly than small lakes, partly because, on average, their sources of sediment are further away. However, this is a low sedimentation rate even for large lakes. Although Lake Hartwell was built upstream on the Savannah River in 1963, and probably reduced the sediment carried into Clarks Hill Lake, it could not have removed the bulk of the sediment. The area draining into Clarks Hill Lake is three times as large as for Lake Hartwell and includes much of the more erodible Piedmont, whereas Lake Hartwell drains mostly forested mountains. Lake Russell was also built upstream in 1983. A statement by Hoke (2000) about the lack of surveys between 1973 and 1999 shows that sedimentation of Clarks Hill Lake must have been low even before 1973; "Because the 1973 survey reflected relatively minor change in the cross-sections, there was a [sic] little interest in resurveying the lines in the intervening 26 years."

Recent concerns have been expressed for pollution of Lake Allatoona, and much of the concern was for sediment deposition. An editorial in the Atlanta Constitution (1999) quoted a recent report as predicting "that Lake Allatoona in Cherokee and Bartow counties will be unfit for water or recreation in just 10 years, largely because of runoff." The US Army Corps of Engineers (2000) surveyed Lake Allatoona three times, in 1949 when the lake was constructed, in the period from 1981 to 1986, and in 1998. In 1998, 108 ranges scattered over the lake were run and compared to the same ranges measured earlier. These surveys show a slow filling of the lake with silt, amounting to a 3.36 percent reduction in the volume of the lake from 1949 to 1998, a rate of only 0.07 percent per year. As with all lakes, the areas where streams enter fill up the most. Where a stream enters a lake, water movement slows and sediment drops to the bottom. So, although the overall sedimentation rate for Lake Allatoona is low, silting has occurred at a higher rate where streams enter. Even though the overall sedimentation was 0.07 percent of the volume per year, the smallest 25 percent of the ranges, which represent mostly shallower areas where streams enter, silted at about 0.5 percent per year.

The long-term sedimentation of Lake Allatoona is low, but it has probably decreased over the fifty-year life of the lake. The average rate from 1949 to 1983 was 0.134 percent per year (Table 3.3), but for the 1983–1998 period the rate was negative (-0.083 percent per year). The negative value for the latter period does not imply an emptying of sediment from the lake, but probably indicates that the surveys were not precise enough to measure the small change that has occurred since the early 1980s. Whatever the precision, it is likely that sedimentation of Lake Allatoona has decreased during the last fifty years because sediment delivered to the lake has decreased. Although measurements of sediment suspended in the Etowah River upstream at Canton are few, they show a downward trend. The average of 204 measurements made by the US Geological Survey in the 1960s and 1970s was 130 milligrams per liter. In the 1990s and 2000 only 20 measurements were made and the average was 13 milligrams per liter. If the sediment in the Etowah River, the main source of water to the lake, is really only 10 percent of that in the 1960s and 1970s, then sedimentation of Lake Allatoona should be less than in the early years. The Lake Allatoona Watershed is 70 percent forest and less than 10 percent agriculture.[69] Likewise, the Coosa River watershed on which the lake is located was estimated in 1990 to have only 6.6 percent of its land cultivated or exposed, [70] so little erosion would be expected.

The case of Lake Jackson, one of the oldest in the state, is a bit more of a puzzle. Lake Jackson was built in 1910, and a survey of the lake in 1935 showed that 12 percent of the volume had been filled with silt, a rate of 0.51 percent per year.[71] A resurvey in 1989 showed that the sedimentation of the reservoir had continued at the same rate since 1935[72] (Table 3.3). There were reservations expressed about the survey, however, that "Much of the morphometric data [measurements of size and shape] for this 90 year-old lake was unavailable or had no defensible reference sources and was therefore incomplete." The concern was the uncertainty about measurements made early in the century. The 1990 report gave the area of the lake as 3,282 acres, although the 1935 survey had shown it as 4,537 acres. Actually the original estimate made by Georgia Power Company was 4,752 acres. So, if the numbers are correct, the lake not only filled with sediment at a rate of 0.51 percent per year for most of the century, but its area shrank by 30 percent. If Lake Jackson filled at the same rate in the last half of the twentieth century as earlier, it must have been largely from sediment stored in stream bottoms rather than new sediment from the uplands. The Georgia Department of Natural Resources study shows that in 1990 the watershed was covered by forest, wetlands, open water, and pastures to the extent of 86 percent and only 4.8 percent of it was cultivated or exposed soil. So it is likely that the erosion from the uplands was small compared to 1936 when the watershed had 40.8 percent cropland and 10.9 percent idle land "usually undergoing serious erosion."[73]

Much of the sediment that is still entering the streams or migrating down the stream channels of the watersheds cannot reach the large reservoirs because of the large number of small reservoirs that have been constructed in the last half of the twentieth century. Although Range (1954) stated that "more than 4,400 farm ponds" had been built by 1949, and the SCS estimated the number at 5,955 in 1950[74], the rate of construction accelerated just after that (Figure 3.7). The Georgia Water Use and Conservation Committee reported in 1955 that there were already 30,000 farm ponds ranging in size from one acre to 200 acres, and 250 new ones were being built per month. If the cumulative numbers for 1950 and 1955 are correct, the rate during that period was about 400 per month. In 1981, the State Soil and Water Conservation Committee reported that there were 79,000 farm ponds occupying 260,000 acres. Another report in 1993 (US Department of Agriculture) gave the number as 63,000 small lakes and ponds covering 312,000 acres. The later number may include only those that USDA aided in construction, but it is clear that there were large numbers of small ponds in Georgia by the 1990s, probably at least an average of 400 to 500 per county. Merrill et al (2001) estimated that there were 5,428 ponds of 25 acres or less in the 1.8 million-acre Upper Oconee watershed. If the abundance over the rest of the state is similar there are about 100,000 small ponds in Georgia. These small reservoirs serve as efficient sediment traps and according to the US

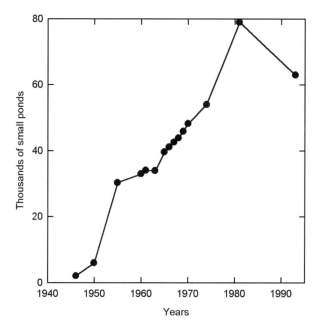

Figure 3.7. Increases in the number of lakes and ponds in Georgia since the middle of the twentieth century. The numbers are from various sources, mostly from reports of the US Department of Agriculture (1960–1970).

Department of Agriculture (1993), "Assuming each pond has a drainage area of 100 acres, they collectively control sediment from approximately 7 million acres."

Concern most often expressed in recent years for erosion and sedimentation is in metropolitan Atlanta and is related to residential and commercial development and the associated construction. According to Soto (1999), "Every day, 600 tons of dirt, the equivalent of 30 dump truck loads, wash into West Point Lake near Lagrange." This sounds like an unacceptable amount of "dirt" but it is a small fraction of the sediment transported by the Chattahoochee River in the early part of the twentieth century. Trimble (1974) reported the "five minute maximum rates" of sediment that were recorded at West Point for four decades before West Point Dam was built. For the decade 1930–1940, the average maximum concentration was approximately 5,000 milligrams per liter. By 1940–1950, it had dropped to 1,000 milligrams per liter and was further reduced to 400–500 milligrams per liter from 1950–1970. The "600 tons of dirt" is only 22 percent of the 2,648 tons per day carried through Atlanta by the Chattahoochee River in the 1930s. A sediment load of 2,648 tons per day was calculated from the average suspended sediment (320 milligrams per liter from Figure 4.1) and the average flow rate of the Chattahoochee River in Atlanta (3,000 cubic feet per second). Sediment is no longer measured at the earlier Atlanta station used since the 1930s and shown in Figure 4.1. If, however, the same calculation is made for the station at Interstate 285 west of the city for the 1990s, the amount of sediment transported through Atlanta is 311 tons per day, 12 percent of that in the 1930s. The amount of "dirt" carried by the Chattahoochee River in the 1930s certainly increased significantly by the time it reached Lagrange because of the streams feeding into it below Atlanta. So, the amount being carried by the Chattahoochee River at Lagrange in the 1930s must have been at least ten times higher than the amount Soto complained about in

1999. Two surveys of West Point Lake by the US Army Corps of Engineers (1998) show that sedimentation has reduced the volume of the lake by 0.34 percent between 1978 and 1997, a rate of only 0.02 percent per year. Compare this to the complete filling of Morgan Falls Lake just above Atlanta in less than thirty years early in the twentieth-century.[75]

Actually, Atlanta is not an erodible landscape in spite of all the publicity about it. The coverage of the urban Atlanta watersheds with buildings and pavement protects the soil from erosion. Even

Table 3.4 The percentage of land covered by forest or cultivated and exposed in the area of metropolitan Atlanta in the 1930s and 1990s. The metropolitan area is defined as the 13-county ozone nonattainment area designated by the Environmental Protection Agency.

County	% cult., exposed 1989-90	% cult. 1934	1997	% forest 1989-90	1934
Clayton	6.5	43.2	30.8	60.9	37.3
Cobb	3.79	37.2	23.0	69.5	36.6
Coweta	3.77	35.0	68.9	81.4	48.1
Dekalb	5.08	31.6	52.6	67.5	39.6
Douglas	4.15	32.9	63.2	79.7	61.7
Fayette	5.38	41.6	47.2	74.3	46.6
Forsyth	6.83	36.3	47.2	68.6	49.2
Fulton	4.18	28.9	37.1	71.4	42.6
Gwinnett	5.0	40.1	37.8	73.0	35.1
Henry	5.87	46.9	53.2	74.9	37.5
Paulding	4.49	31.4	67.9	82.3	54.9
Pike	6.51	44.9	57.9	71.7	29.7
Rockdale	3.87	37.7	47.7	80.0	38.8
Average	5.03	37.5	48.8	74.3	42.9

Acreages for 1989-90 from the Georgia Department of Natural Resources, 1996; for 1934 forest from the U.S. Department of Agriculture, 1946; for 1934 crops from Georgia State Planning Board, 1939; for 1997 forest from Thompson, 1998.

on those areas not covered with buildings, the land is not very subject to erosion. Cultivated cropland in the thirteen-county metro area decreased from 37.5 percent of the landscape in 1934 to only 2.5 percent by 1989. In addition, forest land has increased in the metro area. The US Forest Service estimated that forest cover in the Atlanta metro area was 48.8 percent in 1997 compared to 43.8 percent in 1934 (Table 3.4). However, from satellite images in 1989–1990 it was estimated that 74.3 percent of the metro area was covered by forest and only 5.0 percent by cultivated or exposed soil.[76] Fulton County in the center of this area was estimated to have 76.5 percent coverage of forest plus pasture (includes lawns and grassy playgrounds) and only 4 percent cultivated or exposed earth. Such a landscape is certainly not very conducive to erosion. The 37.5 percent of the area in cultivated crops in 1934 (not counting erodible roads and roadsides) means that there was several times as much exposed land in the early part of the twentieth century.

The extent of land covered by trees is a large factor in the stability of the Atlanta landscape against erosion and there are a lot more trees than some forest estimates show. Discrepancy between the amount of forest in the metro area estimated by satellite imagery in 1989–1990 (74.3 percent) and the Forest Service in 1997 (48.8 percent) is probably due mainly to the definition of "forest land." In the Forest Service surveys forest land must be "Land at least 10 percent stocked by forest trees of any size, or formerly having had such tree cover, and not currently developed for nonforest use. The minimum area considered for classification is 1 acre. Forested strips must be at least 120 feet wide."[77] So there is room for many trees that might show up on satellite images of urban areas but not be counted by the Forest Service as "forest land." In an analysis of the amount of wood in Georgia that might be used for energy, Tansey and Cost (1984) calculated that nonforest land in urban areas in Georgia had an average of 8.8 tons per acre, 12 percent as much as commercial forest.

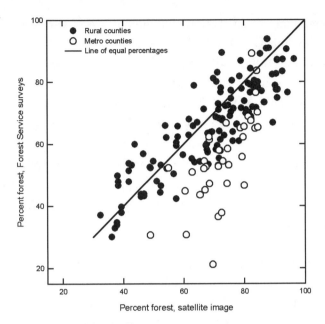

Figure 3.8. Estimates of forest acreage in Georgia counties from satellite images and by US Forest Service surveys. Forest Service estimates from Thompson (1998); estimates from satellite images from the Georgia Department of Natural Resources (1996a).

Rural counties have no more forest cover estimated by satellite images than forest surveys, but urban counties do (Figure 3.8). Counties designated as "Urban statistical areas"[78] had more forest by the satellite image estimates (they fall below the line in Figure 3.8), probably because the Forest Service definition of "forest land" leaves out so many trees. The largest discrepancies (the points to the lower right in Figure 3.8) were for five counties in the center of the Atlanta metro area, Clayton, Cobb, Dekalb, Gwinnett, and Fulton. The differences between the 1989–1990 and 1997 estimates are not likely to be due to the removal of trees in the metro area, although there have been several references to the amount of forest lost in the 1990s (for example, 50 acres per day[79]). The estimates by satellite imagery that trees cover 60–70 percent of the land in these five counties, along with the estimates of very small areas of exposed soil, if accurate, mean that there is little production of sediment from lands in the metro area to foul the streams.

In a study of the amounts and sources of sediment for the Chattahoochee River, Faye et al (1980) estimated that erosion was least in the watersheds of Peachtree and Nancy Creeks, the most urbanized in the Chattahoochee River Basin. The average of estimates for these watersheds was 975 tons of soil per square mile per year compared to 6,390 tons per square mile for the relatively undeveloped Soque River watershed in the Upper Chattahoochee River basin. Even the forested watershed of Snake Creek south of Atlanta had twice the erosion of the urban creek watersheds in Atlanta. Faye et al considered a large part of the sediment discharged in urban streams to be derived from erosion of stream channels. The SCS seemed to welcome urbanization as a means of reducing sediment pollution of streams and lakes. The following is from a 1961 report on sediment control in Georgia: "Urban expansion and development, particularly in upland areas of the Piedmont, should result in greatly reduced rates of erosion and sediment production for lands so affected."[80] The average urban erosion rate for Georgia was listed as zero in 1992 by the US Department of Agriculture (1994) compared to 5.5 tons per acre for cultivated crops (Table 3.1). So in spite of the construction in urban areas of the Piedmont, the amount of soil washed off the land must be a small fraction of that early in the twentieth century.

Urbanization certainly does affect streams, however. According to Ferguson and Suckling (1990), the covering of so much of the watershed with impervious surfaces deprives the streams of water during dry periods because groundwater is not recharged during wet weather. On the other hand, the buildings and paved streets and parking lots divert water to streams faster during heavy rain and cause more severe flooding and erosion of the stream channels. The flooding is often made worse by bridge abutments, pipelines, and other structures that cross the streams and partially obstruct flow of water. Contrary to the countryside, in a heavily urbanized area, the effects of water are shifted from erosion of uplands and silting of streams to erosion of the stream channels as Faye et al (1980) suggested for Peachtree and Nancy Creeks. Too much building in flood plain areas usually worsens the property damage caused by urban flooding.

Major changes in the landscape are almost always a mixture of blessing and misfortune. It is true for the countryside as well. One of the consequences of the reforestation of the countryside is that it has reduced the flow of many streams. When rain falls on a watershed it either 1) evaporates from the soil and vegetation or 2) drains to streams or underground reservoirs. Bare soil evaporates very little water except when the surface is wet. However, much more water evaporates when vegetation covers the ground. For this reason, farmers in dry regions maintain fallow fields to keep weeds from growing and wasting water in years when a crop is not being grown. Evaporation of water, even that deep in the soil, occurs if the watershed is covered with deep-rooted trees. So reforestation of the watersheds in Georgia has led to more of the water evaporating and less of it running in the streams, compared to earlier times when the large row-crop acreage was bare for much of the year. Trimble et al (1987) surveyed the river basins in the Southern Piedmont and found that reforestation of seven major river watersheds in Georgia during the period 1919 to 1967 ranged from 11 to 27 percent. For the streams draining those watersheds flows were reduced 4 to 21 percent. The average net loss of water due to reforestation, that is, the increase in the amount evaporated, was equal to about 12 inches of rain. Bartlett (1993) notes that streams in Dodge County went dry in the droughts of the 1980s that had not dried up in the memories of men farming the land. The reason may be the reforestation of 26 percent of Dodge County, from 37 percent of the land in 1935 to 63 percent in 1989.

In spite of the low erosion and the resulting reduction of silting of lakes and ponds, the success of soil conservation is not well known among the public. One reason is that the SCS is an action agency. It responds to perceived needs, writes farm plans, formulates and coordinates watershed plans, and promotes soil and water conservation. It and the local Soil and Water Conservation Districts continue to indicate the need for improved soil conservation, rather than emphasizing the improvements. In 1981, the State Soil and Water Conservation Committee, which represents the districts at the state level noted hardly any successes in its "Georgia Resource Conservation Program and Action Plan," but instead emphasized the problems. The following statement illustrates the point about cropland; "Erosion of cropland, once largely under control, has become one of the most critical resource problems confronting the state."

This contrasts with the following statement of Jimmy Carter nearly ten years earlier at a conference he convened on sedimentation. "I have been grieved, since becoming governor, to see that the tremendous progress made in erosion control in rural areas has almost been offset by the worsening circumstance of erosion and sedimentation in the urban areas of Georgia."[81] Governor Carter knew well the conservation efforts, having been active in agriculture for many years. In the years between the governor's conference and the action plan, cultivated crops did increase by about 1 million acres, but it was a temporary increase (see Figure 2.1) and probably affected statewide erosion to a minor degree. In the same 1981 publication, the State Soil and Water Conservation Committee reported that "excessive" erosion was occurring on 4 million acres in 1979, even though that was about 90 percent of the total acres planted to row crops that year. It is unlikely that 90 percent of the land in row crops had "excessive" erosion, particularly since the lands most susceptible to erosion were no longer being cultivated and conservation methods were in wide use. The definition of "excessive" may be the key, however. If soil erodes at a rate higher than it is being formed, it might be considered "excessive." The rate of soil formation is considered to be about 4.5 tons per acre per year for Georgia, about a ton per acre lower than the present erosion rate on cultivated land.

The concern about erosion control did not fade in the twentieth century; the problems continued to be stressed to the end. After saying in 1973 that a good deal of progress had been made in air pollution, strip mining, and water pollution control, Governor Jimmy Carter concluded that, "Of all the problems still not faced, sedimentation is most important." During Carter's administration, the Erosion and Sedimentation Act was passed for the establishment and implementation of a comprehensive soil erosion and sediment control program.[82] A federal Soil and Water Resources Conservation Act was passed in 1977, and in surveys resulting from this act, erosion was listed by more of the 4,000 Georgia respondents as the top priority than any of the other twenty-one issues.[83] It is likely that a large percentage of the respondents were farmers or agricultural business people who had an interest in farming.

However, soil erosion even came to be looked on as an urban problem. An *Atlanta Constitution* editorial said at the end of the century (1999), "State environmental-protection officials admit that the most serious water-quality problem facing Georgia is soil eroding from our many construction sites into streams and lakes." Environmental action groups like the Upper Chattahoochee Riverkeeper continued to advocate erosion control, mostly from construction sites in urban and suburban areas. It is certain that soil erosion became a minor problem late in the twentieth century compared to that in the early years, but because environmental problems are relative, the solution of so many other water pollution problems made soil erosion and stream sedimentation again the most serious. And the problems of the country came to town.

How Did It Happen?

However, as I have tried to show in this chapter, soil erosion in Georgia is at its lowest in at least 100 years. The main circumstances that led to reduced soil erosion in Georgia are; 1) the reduction in acreage of row crops, 2) the shifting of row-crop agriculture to South Georgia, and 3) the increased use of soil conservation practices. The decrease of row-crop acreage and its replacement with less erosive uses has been described in chapter 2. These first two circumstances are largely the result of forces other than those directed at saving the soil. Although the Soil Bank Program and the Resource Conservation Act helped reduce erosion by reducing crop acreage, their aim was as much to stabilize farm income and help protect the agricultural economy as to protect the soil. Rental payments for land retired for conservation purposes became the largest category of US Department of Agriculture conservation expense in the 1990s, exceeding $3.9 billion in 1993[84]. Held and Clawson (1965) claim that even the funding of early soil conservation efforts (the shift from "soil-depleting" to "soil conserving" crops) by the Agricultural Conservation Program was intended as economic aid to farmers. They quote others as suggesting that "the purposes of the program were deliberately confused: presented as a soil-conservation program, in reality it was an income-assisting program."

The greatest improvement in agriculture that allowed the land to be restored is the increased productivity of the land of Georgia and other states, that is, the increased efficiency means we no longer have to grow crops on the most erodible land. According to Odum and Turner (1987), "...in the present decade (1980s), Georgia grows twice as much food on half as much land as in the 1934–44 decade." "Net primary productivity" (yields of the whole plant) increased over 350 percent for cotton, 605 percent for corn and 400 percent for peanut.[85] Increased productivity was due to a variety of changes, such as increased use of fertilizers and pesticides, genetically improved crops, and abandonment of poorer soils. Although much of the increased productivity of crops is due to use of fertilizers and pesticides, cited by many as major causes of pollution, a doubling of forest productivity from 3 to 6 tons per acre from 1935 to 1982[86] involved little use of fertilizers or pesticides. Although the increased productivity of forest is likely due mostly to improved management, such increases in productivity are indicative of an improved environment, if for no other areas than the forests themselves.

The improvement of the economy since World War II is another main reason for the restoration of Georgia's lands. A major cause of land mismanagement is poverty. Perhaps not the kind of poverty defined by family income below a certain dollar level, but malnutrition, squalid housing, poor hygiene, the numbness and hopelessness of the human spirit that is tied to a poor land ethic. Albert Cowdrey states in his book that poverty is no friend to natural resources which "typically are devoured piecemeal to sustain existence."[87] In a report of the Council on Environmental Quality (1994), President Clinton is quoted as saying "The fact is that only a prosperous society can have the confidence and the means to protect its environment." The countries of the world where people live in ignorance and deprivation are, even today, the ones where land conservation is neglected most. The southern United States, including Georgia, has a long history of land mismanagement. It is probably not a coincidence that the Southern states also had the poorest, least educated population in the nation.

The changes in the last half of the twentieth century have erased much of the regional stigma. Visions of shabby sharecropper houses, rutted, eroded country roads, and barren red hills are gone, except in books and the faded memories of older Georgians. If Sydney Lanier could return, he would know that today's Jones County farmer (the few that are left; Jones County had 84 percent of its land in forest in 1997) has no "itch" to "git places in Texas." He would see that the "mouldering mill" below the scarified old

Georgian hill is no longer needed, but both the hill and the stream are largely repaired. Henry Grady's expectations of the "New South" would be more than fulfilled. The South has probably changed more than most of the country because of the southward movement of industry and commerce. It is no accident that the greening of Georgia has come in an era of unprecedented economic and social prosperity. The prosperity has allowed time for concern about many human problems beyond the immediate needs of food, clothing, and housing. The enlightenment of the public about environmental conditions and an unprecedented advance in technology have allowed an environmental revolution. This quiet revolution includes a revised ethic that has allowed a restoration of the land. Dr. Bennett would know that his cause has been met.

The dream of that gentle crusader of an earlier generation, Sidney Lanier, tendered at the end of "Corn" is prophetic.

> Old hill! old hill! thou gashed and hairy Lear
> Whom the divine Cordelia of the year,
> E'en pitying Spring, will vainly strive to cheer—
> King, that no subject man nor beast may own,
> Discrowned, undaughtered and alone—
> Yet shall the great God turn thy fate,
> And bring thee back into thy monarch state
> And majesty immaculate.
> Lo, through hot waverings of the August morn,
> Thou givest from thy vasty sides forlorn
> Visions of golden treasuries of corn—
> Ripe largesse lingering for some bolder heart
> That manfully shall take thy part,
> And tend thee,
> And defend thee,
> With antique sinew and with modern art.

Though Sidney Lanier could not have imagined the kind or extent of "modern art" now employed, the restoration of the "old hills" is in large measure due to the application of modern scientific methods for production of food and management of the land.

[1] Evans et al, 1913
[2] Trimble, 1974.
[3] See chapter 2.
[4] Aderhold, 1962.
[5] Trimble, 1974.
[6] Sir Charles Lyzell, A Second Visit to the United States (1849), quoted in Trimble, 1974.
[7] Trimble, 1974.
[8] Hilgard, 1884.
[9] Trimble, 1985.
[10] Coulter, 1965.
[11] Lanier, 1981.
[12] Jackson, 1850.
[13] Tate, 1960.
[14] Trimble, 1974.
[15] Eakin, 1939.
[16] Works Progress Administration, 1990.
[17] Magilligan and Stamp, 1997.
[18] Langdale et al, 1979.
[19] Trimble, 1974.
[20] Trimble, 1970.
[21] Bagby, 1969; Wharton, 1970.
[22] Bagby, 1969.
[23] For one such objection, see Bagby, 1969. Georgia Game and Fish, 1969.
[24] Georgia Department of Natural Resources, 1990b.
[25] Ferguson, 1997b.
[26] Chapman et al, 1950.
[27] Broad River Soil and Water Conservation District et al, 1969, 1974.
[28] See later in this chapter.
[29] Georgia Department of Agriculture, 1901.
[30] Eakin, 1939.
[31] Albert and Spector, 1955.
[32] Burleigh, 1937.
[33] Sisk, 1975.
[34] Works Progress Administration, 1990.
[35] Sisk, 1975.
[36] Range, 1954.
[37] Sisk, 1975.
[38] Sisk, 1975.
[39] Sisk, 1975.
[40] Sisk, 1975.
[41] Sisk, 1975.
[42] US Department of Agriculture, 1960–1970.

[43] Held and Clawson, 1965.
[44] Broad River Soil and Water Conservation District et al, 1969, 1974.
[45] Trimble, 1985.
[46] US Department of Agriculture, 1977.
[47] Trimble and Crosson, 2000.
[48] Magilligan and Stamp, 1997.
[49] See chapter 2.
[50] Georgia State Soil and Water Conservation Needs Inventory Committee, 1962.
[51] State Soil and Water Conservation Commission, 1994
[52] State Soil and Water Conservation Commission, 1994.
[53] Citizens' Fact-Finding Movement of Georgia, 1946.
[54] Range, 1954.
[55] Chapman et al, 1950.
[56] Richardson and Diseker, 1961.
[57] State Soil and Water Conservation Committee, 1981.
[58] Chapman et al, 1950.
[59] Trimble, 1975.
[60] Oppenheim, 1996.
[61] Craft and Casey, 1999.
[62] Magilligan and Stamp, 1997.
[63] Burke, 1996.
[64] Ferguson, 1997b.
[65] Meade, 1982.
[66] Spraberry, 1965.
[67] Bates et al, 1988.
[68] Hoke, 2000.
[69] Burruss Institute of Public Service, 1999.
[70] Georgia Department of Natural Resources, 1997.
[71] Eakin, 1939.
[72] Georgia Department of Natural Resources, 1990b.
[73] Eakin, 1939.
[74] US Department of Agriculture, 1950.
[75] Eakin, 1939.
[76] Georgia Department of Natural Resources, 1996a.
[77] Thompson, 1998.
[78] Kundell, 1996.
[79] Seabrook, 2000b.
[80] US Department of Agriculture, 1961b.
[81] Carter, 1973.
[82] Georgia Department of Natural Resources, 1976.
[83] State Soil and Water Conservation Committee, 1981.
[84] Council on Environmental Quality, 1994.
[85] Turner, 1987.
[86] Odum and Turner, 1987.
[87] Quoted by Odum and Turner, 1987.

CHAPTER FOUR

Cleaning of the Water

"The poet Sidney Lanier, who glorified the river in his *Song of the Chattahoochee* in 1877, would once again be proud of his river"[1]

Water Quality in the Past

The erosion of the uplands in the 1800s and early 1900s probably limited Georgia's future progress more than anyone realized, either then or now. It certainly reduced productivity of the land in the last two centuries and still does. But it also contaminated the streams with silt and clay. The few records of water pollution in the first half of the twentieth century show high concentrations of soil particles in the streams, especially in the Piedmont. The Chattahoochee River in Atlanta, for example, had 300 to 400 milligrams per liter of suspended sediment in the 1930s (Figure 4.1). The muddy water was unfit for drinking and hygienic purposes without cleaning and must have caused many engineering problems for industries that used the water. Muddy water also caused harm to the fish population of the streams. I remember in the 1940s when Buckhorn Creek on our farm quit running in the summer. The fish were confined to the deeper holes that still contained water and we would stir up the mud in the water and wait for the fish to rise. Because they could not tolerate the muddy water, they came up to the surface and were then easier to catch by hand or with a dip net.

Washing of soil into the streams was the first and most serious pollution that occurred in Georgia, but it was not the last. Local fouling of streams by animal manure, human wastes, and effluent from small industries, such as tanneries, occurred soon after the settlement of Georgia. It is likely that parts of rivers and streams near cities and towns were highly polluted since there was no treatment of human waste before release into the streams. Local pollution of small streams by the washing of animal waste from farmyards was very likely. There is little evidence of the pollution of water by animal and human waste before the mid-1900s, however, because there was no analysis of water then. But there were references to poor

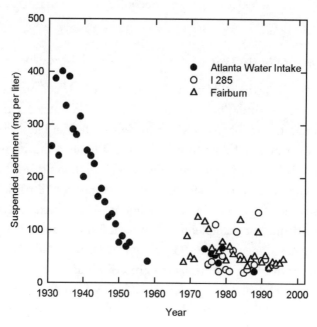

Figure 4.1. The trend of suspended sediment in the Chattahoochee River near Atlanta. Data are from Albert and Spector (1955) and records of the US Geological Survey.

water quality as far back as 1892, when the Street Committee of Dublin, Georgia reported that ditches and drains were in such poor condition that "we consider the health of the entire city endangered." In the early 1880s, in Hawkinsville, "Garbage, rotting vegetation, and trash blocked the drainage ditches, which stank from the waste of dozens of privies."[2] Hawkinsville and Dublin were relatively small towns then and the problems must have been greatly magnified in the larger towns of Georgia.

Safety of drinking water supplies was seldom questioned until there was an outbreak of disease. Among the diseases that caused death early in this century, several were associated with poor water quality. Typhoid, dysentery, diarrhea, and malaria together accounted for 8.8 percent of deaths in 1920, but only 1.2 percent in 1949, and after mid-century they were not even listed among causes of death.[3] Poor sanitary conditions in Atlanta early in the twentieth century were linked to disease. According to Moran (1995), "Malaria and polio were ever present in Atlanta; cases of malaria increased from 3,972 in 1935 to 11,408 in 1936. The city was listed in the top ten for cases of diphtheria and typhoid, and 15 cases of smallpox appeared statewide in 1938. The outbreaks were attributed to the poor sewage system and separate city and county health departments which were inefficient and frequently duplicated services." If as many cases of an environmentally related disease occurred in Atlanta today as there was of Malaria in 1936 it would be considered a catastrophe, even though the current population is nearly four times greater than in 1936.

Therefore, it is surely wrong to assume that water was universally clean in the late 1800s and early 1900s, because many of the local conditions we now know contribute to contamination of water certainly existed. The large proportion of the population that lived on the farm was likely exposed to polluted water more often than anyone knew. The typical farmstead before World War II had outdoor toilets and livestock pens close to the dwelling. As late as 1955, more than half of Georgia's rural families did not have an indoor water supply. Wastewater from the kitchen or home laundry was simply thrown into the yard or garden in most cases. Left-over food was fed to dogs or swine and solid waste (paper, etc.) was usually burned in a container or in the open. The well for drinking water was usually located very close to the dwelling and in many cases was less than 50 feet deep. The potential for seepage from livestock pens, toilets, and household waste into shallow wells was great. But we could hardly have known the threat. I remember that we considered contamination of our shallow well to occur when a small animal got trapped and died in it, or when mosquito larvae showed up in it during dry summers.

The drinking water in towns and cities was unlikely to have been much safer. Urban areas disposed of their household wastes in open untreated dumps, usually without regard to proximity of streams or underground drinking water sources. One source of pollution of streams was the dumping of human wastes into streams mainly through sewers. Some larger towns had sewage disposal systems but none treated the sewage before release into nearby streams. Sewers for removal of human wastes from homes were not used until late in the nineteenth century, even in large cities, and technology to treat sewage was not available until the early 1900s. Atlanta had only 24 miles of sewers in 1890 and 122 miles in 1905[4] when its population was about 90,000. Wastes in most cities were disposed of in the same way as in rural areas until sewage treatment methods were developed, but the concentration of people and commercial development in cities compounded the problem of disposing of waste. It was mid-century and later before much improvement in sewage treatment occurred. In 1958, the US Public Health Service listed 262 sources of pollution in Georgia "which are of significance in water pollution control programs." Of these sources, 240 were municipal sewage systems.[5] As late as 1964, 70 percent of all the municipal sewage and 97 percent of all industrial wastewater in Georgia was discharged without treatment.[6]

Concentration of large numbers of animals in livestock feeding operations is currently a matter of concern, but the concentration of animals in cities early in the twentieth century must have been a far worse problem. Traffic in Georgia towns and cities before World War II was mostly by animal power and the problem of animal waste in the urban areas was one of considerable concern, but its effect on human health was probably underestimated. Dust and flies were common complaints, but contamination of water supplies was unappreciated, because of the lack of awareness and technology. Most streets in most towns were unpaved. The manure from draft animals accumulated in the streets, and washed into streams along with mud and other street debris when it rained. There were even dead animals to worry about as indicated by this firsthand account from an Atlantan early in the twentieth century: "City and county officials employed wagons to pick up "night soil" from outdoor privies, and worried about the disposal of dead animals."[7] In the late 1890s, insurance company actuaries discovered that employees in livery stables and those living near stables had a higher rate of infectious diseases, such as typhoid fever, than did the general public. The smell of horse manure in city streets, and the dust from dried manure, ground to a powder by hooves and wheels, were a nuisance and a health hazard.[8] Hogs were so numerous in Eastman, Georgia in the early twentieth century that one resident declared "Hogville" a more appropriate name for the place and composed a ditty to characterize the situation: "Piggies, piggies on the street, All around you nice and sweet,/When we go our friends to meet, Piggies, piggies we will greet."[9] There were about 45,000 horses and mules and over 100,000 hogs in Georgia towns and cities in 1920,[10] and it is certain they contributed to the unhealthy conditions frequently described.

The virtues of wetlands have been extolled in recent years, but in the early part of this century, they were described as swamps and were considered useless and worse and in need of draining to provide a healthier environment and productive farmland. Swamps were considered to be sources of dreaded diseases, including malaria and typhoid. In a history of Macon, Young et al (1950) stated that, "During 1907 Dempsey Pond at the corner of Seventh and Walnut Streets was drained as a health measure..." The source of health concern for the pond may have been its polluted condition. The Soil Conservation Service and the Army Corps of Engineers spent considerable sums to channelize streams (drain swamps) in the Piedmont that had been filled by erosion sediment, in addition to some in South Georgia that that were deepened to drain natural swamps.

The good news about water in Georgia is that it is cleaner now than earlier in the twentieth century. To present that good news, I have examined records of water quality for wells and the major streams in the state, and will use mainly those records that are most complete and extend back farthest in time. Most of the records are from measurements made and/or published by the United States Geological Survey and the Georgia Department of Natural Resources (and a predecessor, the Georgia Water Quality Control Board). Except for suspended sediment, most of the records do not go back farther than the late 1960s or early 1970s. Because of the large variation in some data from year to year, three-year running averages have been plotted in some figures to smooth the trends.

Clearing the Streams

The muddy appearance of streams is due to particles suspended in the water. The amount of solid particles suspended in streams is usually expressed as milligrams per liter or the equivalent measure, parts per million. This suspended sediment is mostly soil particles eroded into the streams by rainwater or picked up from the stream channel by the turbulent action of the water. The pollution of streams by erosion has been minimized since the early part of the twentieth century. Since soil erosion is a natural process, it will never be completely eliminated, but erosion in Georgia today is a small fraction of what it was early in the twentieth century. As a result of the decrease in soil erosion in the century the amount of sediment suspended in the Chattahoochee River at Atlanta is less than 50 milligrams per liter, compared to 300 to 400 milligrams per liter in the 1930s (Figure 4.1). Sediment in streams is usually higher in summer, and according to Trimble (1974), the August concentrations for Atlanta dropped from 800 milligrams per liter in 1931 to less than 100 milligrams per liter in 1952. A similar drop in sediment has occurred in the Ocmulgee River at Macon, the Savannah River at Augusta, and the Oconee at Milledgeville.

The sediment in the Chattahoochee River in recent years is mainly from two sources; land disturbance by construction in the Atlanta metropolitan area and re-suspension of sediment from the riverbed that washed from the uplands in prior decades. The large variation in suspended sediment from the 1970s through the 1990s at Interstate 285 west of the city and Fairburn 20 miles to the south (Figure 4.1) may reflect changing construction activity in various areas of metropolitan Atlanta and variability in sampling. If sampled after a heavy rain, sediment in the water is always much higher. Only two annual averages for the Interstate 285 station were above 100 milligrams per liter and one high monthly reading each year caused those high values. If the highest readings are left out, the averages for those two years (1977 and 1989) are below 50 milligrams per liter. Excluding those two years, the annual average for the Interstate 285 location from 1975 to 1994 was 40 milligrams per liter, similar to that for Fairburn in the 1990s.

Sedimentation of the Chattahoochee is still a matter of concern. A recent article in the *Atlanta Journal-Constitution* stated that "Erosion and sedimentation from construction sites are major causes of pollution in the Chattahoochee and its tributaries in metro Atlanta. Soil from erosion can choke the life out of waterways, foul drinking water and make streams unfit for recreation."[11] The article was about the lack of enforcement of erosion control ordinances in Atlanta. In spite of lax enforcement of sediment control laws, if that is the case, and in spite of the land disturbance by construction in Atlanta, the sediment load of the Chattahoochee in the 1990s is less than 20 percent of that in the 1930s. Apparently the modern urban influence cannot match the extensive cultivation of the North Georgia hills in muddying the Chattahoochee.

The muddy appearance of many Piedmont streams today, especially the larger ones after heavy rains, results mostly from resuspension of sediment that is already in the stream beds and banks.[12] Flood plains and uplands have been largely stabilized in the latter half of the twentieth century and stream channels are being continually deepened by erosion of the stream bed itself. This cutting of channels through the old sediment deposited over the last two centuries will likely continue until they are worn down to the original rocky bottoms, which has already occurred in some smaller streams.[13] Soil from the steep banks composed of old sediment on these recently deepened Piedmont streams is washed into the streams when flooding occurs. So the streams become muddy without much soil washing from the uplands.

The streams of the Coastal Plain do not support a high sediment load because less sediment enters the streams and the water moves more slowly. The faster flow of Piedmont streams allows them to carry about ten times as much silt as Coastal Plain streams of the same volume. As a result of the more gentle slopes and slower flowing streams in the Coastal Plain, sedimentation of the streams is less. It has been estimated that the Coastal Plain portion of the Savannah River basin contributes a minor portion of the sediment in the Savannah river, even though it constitutes about one-third of the drainage area.[14] Because of their slower flow,

Coastal Plain streams run clearer than Piedmont streams, although many Coastal Plain streams are dark in color from natural organic compounds. However, even Coastal Plain streams have cleared up in the latter half of the twentieth century. Sediment in the Savannah River near Clyo (about 30 miles north of Savannah) decreased from an average of 82.5 milligrams per liter in 1938 and 1939 to 30.5 in the 1960s and to 15.7 milligrams per liter in the 1990s.

The relative clarity of Georgia's streams today is due in large measure to the reduction of soil erosion resulting from fewer acres under cultivation, and better erosion control practices. As described in chapter 2, row crop acreage has decreased by about 70 percent since 1910. The bulk of the land area in the highly erodible Piedmont is in forest and pastures, which minimizes erosion. The main area of row crop cultivation has shifted from the Piedmont to the Coastal Plain where the land is not as sloping and, therefore, is less susceptible to erosion. About 42.5 percent of cultivated acreage was above the fall line in 1944, but only 17 percent in 1987. In addition, much better conservation practices are used in crop cultivation than early in the twentieth century. The use of minimum tillage or no tillage (no-till), in which the soil remains covered with the residue of the previous crop, has increased since the 1950s.

Another development that has contributed to the clarity of water in the last half of the twentieth century, is the construction of ponds and lakes in all sections of Georgia. In 1955, there were already 30,000 farm ponds ranging in size from 1 acre to 200 acres, and 250 new ones were being built per month.[15] In 1993, The US Department of Agriculture estimated that there were 63,000 small lakes and ponds covering 312,000 acres. These reservoirs plus the dozens of large lakes serve as settling basins for the sediment that enters them. Meade (1976) estimated that a reservoir large enough to hold 10 percent of the water that flows into it in a year, can trap 85 percent of the sediment. Meade further stated that Lake Hartwell on the Savannah River is large enough to hold all of the water that flows into it in an average year. According to Rasmussen et al (1998), it requires about one and a half years on average for water to move through Lake Lanier. It follows that the water flowing from Lakes Hartwell and Lanier is nearly clear of sediment. Therefore, it is easy to see the impact of ponds and lakes on the filtering of the streams. Of course, if erosion were as severe as early in the twentieth century, these water reservoirs would be filling with silt at a rapid rate, as ponds and lakes were in the 1920s and 1930s. However, there is no evidence that silting of reservoirs is now a serious problem in Georgia, in spite of the recent rash of stories in the press about construction-related sedimentation of the Chattahoochee.[16]

Pollution by soil is objectionable and results in an unappealing appearance of the streams, but it is not the most dangerous form of water pollution. The fouling of streams with human, animal, and industrial wastes is more problematic, especially for human health. Industrial development and concentration of the population in urban centers during the twentieth century increased the release of pollutants into Georgia streams. There were a number of chemical and biological agents that posed considerable danger to the health of humans, and to fish and other aquatic life. One source of danger was the contamination of streams with untreated sewage.

Reducing Sewage Pollution

Widespread and massive pollution of streams in Georgia came to public awareness in the 1960s. The State Water Quality Control Board initiated a pollution abatement program on the Chattahoochee River at a conference in Atlanta on 25 February 1965. Analysis of many Georgia streams was conducted in the 1960s to document pollution that many citizens knew of, or suspected. The impetus for most of these analyses was foul odors and visible signs of pollution. Small creeks carrying raw sewage were sure signs of unhealthy water, and large concentrations of people and industries on the larger rivers made it very likely that communities downstream were in danger. One of the signs that a stream is contaminated with sewage is the presence of fecal coliform bacteria. Bacteria of the coliform group inhabit the intestinal tract of animals, including humans, and are used as a test of the presence of human or animal wastes in streams and lakes. Small samples of water are put into artificial culture and the numbers of bacterial colonies that develop are counted. The counts are reported as most probable numbers (MPN).

Routine tests of Georgia streams for fecal coliform bacteria began in the late 1960s. They were found in some streams in large numbers, mainly downstream of towns or cities, and were the result of release of untreated sewage into the streams. A striking example of the location of this problem in the Chattahoochee River in 1970 is shown in Figure 4.2A. The fecal coliform counts were near zero all along the river, except at Atlanta and Columbus. Near Atlanta, the fecal coliform counts jumped to about 1 million per 100 milliliters, and near Columbus to over 500,000. The fact that the fecal coliform counts dropped to very low levels below Atlanta and Columbus, indicates that these bacteria were greatly diluted, settled out of the water, or were otherwise assimilated. In the case of the massive pollution of the Chattahoochee River, however, contamination remained for considerable distances. In both 1968 and 1969, the average fecal coliform counts for the first 14 miles of the Chattahoochee River below Atlanta exceeded 1 million per 100 milliliters. Mr. R. S. "Rock" Howard stated at a conference on pollution of the Chattahoochee River in 1965, that "the Chattahoochee is grossly polluted for about 100 miles below Atlanta from sewage and industrial wastes."[17]

Since the late 1960s and early 1970s, improved sewage treatment has greatly decreased fecal pollution of most of Georgia's streams. Four examples of the trends in fecal coliform contamination are shown in Figure 4.3. The decrease in fecal coliform counts

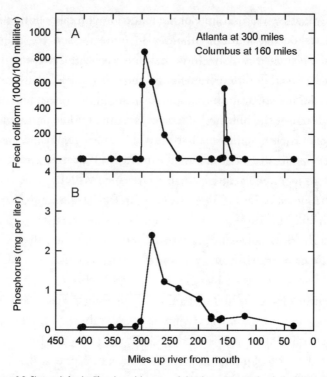

Figure 4.2. Changes in fecal coliform bacterial counts and phosphorus concentrations in the Chattahoochee River with distance from the mouth in 1970. Georgia Water Quality Control Board (1971).

in the Chattahoochee downstream of Atlanta at Fairburn shows the remarkable improvement in sewage treatment in Atlanta. In fact, the decrease in fecal coliform bacteria below Atlanta may have been more than is shown in Figure 4.3. In a conference on pollution of the Chattahoochee River held in Atlanta in 1970, the average fecal coliform bacteria counts at Fairburn were given as between 800,000 and 1,000,000 per 100 milliliters for 1968 and 1969, four or five times the average from US Geological Survey data shown in Figure 4.3A.[18] At that conference, J. L. Ledbetter stated that, "In the first 14 miles below Atlanta, fecal coliform concentrations in the river resemble those found in primary treatment plant effluents." Other streams that show similar improvement since the 1970s, are the Coosa River at Rome, the Conasauga River at Tilton, and the

Flint River at Newton. In spite of the reduction of fecal pollution in the Chattahoochee south of Atlanta, the counts are still not as low as in the Savannah near Clyo (note scale for Peachtree Creek and the Savannah River is different than for the Chattahoochee River). River locations upstream from large towns or cities were not badly polluted with fecal coliform bacteria even in the middle part of the century. However, some small streams in heavily populated areas still show signs of pollution, as illustrated by Peachtree Creek in Atlanta. In fact, improvement there has been only about 30 percent from the 1970s to the 1990s (averages between 10,000 and 20,000 per 100 milliliters).

Another example of extreme fecal coliform pollution is the South River, which rises near downtown Atlanta and flows south-

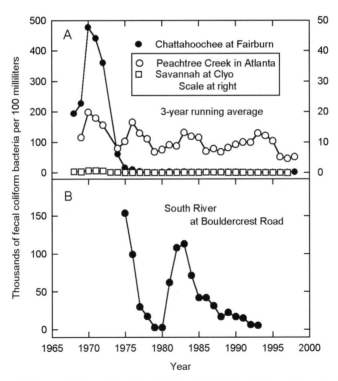

Figure 4.3. Changes with time in fecal coliform bacterial counts in four Georgia streams. Data are from the US Geological Survey records.

eastward to Lake Jackson at the junction of Newton, Butts, and Jasper Counties. This was one of the first rivers in Georgia that suffered massive pollution because of the dense population along its headwaters, and the lack of any treatment of the sewage dumped in it. As early as 1880 it was lined with rocks and covered because of its foul odor.[19] With the advent of sewage treatment, there was improvement in the river, but the growing population and industry overwhelmed the small size of the river. In the 1980s the fecal coliform levels in the South River reached about fifty times the levels in the Chattahoochee at Fairburn at that time. From the mid-1980s to 1990, fecal coliform contamination was greatly decreased because of improved sewage treatment. The levels in the South River in the early 1990s, however, were still higher than most other streams.

As shown by the Peachtree Creek and South River examples, the complete elimination of contamination by human and animal waste appears to be difficult. In the Georgia Nonpoint Source Management Plan of 1989, it was estimated that 97 percent of the 19,443 miles of Georgia streams and rivers supported designated uses, and 2 percent partially supported such usage. However, all eighteen segments (192 miles of streams) which did not support designated use were designated for fishing, and the criterion violated was fecal coliform from urban sources. Fecal coliform or other bacteria are unlikely to reach public water supplies because water is treated before it enters the systems. There is concern, however, about human contact in streams and swimming pools containing fecal bacteria. The state maximum for fishing and swimming waters is 200 counts per 100 milliliters.

As late as 1992, Lake Acworth at Acworth, Georgia, was closed when it was discovered that fecal coliform levels were up to 940 bacteria per 100 milliliters.[20] The source of bacterial contamination was initially linked to a leaking sewer line. However, the continued high levels of fecal coliform bacteria in the lake in 1993 after sewer line leaks were either fixed or ruled out suggested that other nonpoint sources of contamination were involved. It was finally determined that geese and ducks that used the lake were the source of contamination. Relocation of 200 waterfowl by University of Georgia biologists reduced fecal coliform bacteria to "acceptable levels." Thus in this case, wild birds, or more likely "semi-wild" birds, turned out to be a source of pollution.

The dumping of waste in streams causes other problems than those of disease. Any organic matter put into streams requires oxygen for decomposition, and oxygen used for decomposition lowers the amount available to fish and other aquatic organisms. The easier it is for the organic matter to decompose, the faster the oxygen in the water is depleted. Items like food scraps, animal wastes, and organic industrial chemicals lower oxygen more rapidly than slowly decomposing substances such as tree branches and leaves. Large amounts of organic waste in streams can cause so much of the oxygen to be used that fish and other aquatic creatures cannot get enough oxygen to survive. Most fish require a minimum of about 4.0 milligrams of oxygen per liter of water for survival. Trout require higher levels of oxygen, and for a stream to be designated as a trout stream it must have at least 6.0 milligrams per liter. In the past, large stretches of rivers and smaller streams in Georgia were so polluted with waste that they supported few fish, or only the most resistant species. One of the major reasons for the lack of fish was low oxygen concentration.

There are two measurements commonly made that indicate the oxygen supply in streams. One is the direct measurement of dissolved oxygen, usually expressed as milligrams per liter. The other measurement that gives an indication of the potential oxygen problem in a stream is the biochemical oxygen demand (BOD). This is a measure of the decomposable organic matter in a stream and of the oxygen required to break it down. There are a number of things that affect the amount of oxygen in streams, some of which have nothing to do with pollution. Warm water holds less oxygen than cool water, and so North Georgia streams naturally have more oxygen than South Georgia streams because they are cooler. North Georgia streams are also likely to have more oxygen because they are more aerated by the tumbling and mixing of rapid flow compared to the slow moving streams in the south. In addition, when water becomes still, as it does in large lakes and ponds, the bottom water contains less oxygen because oxygen from air cannot diffuse rapidly through thick layers of still water. Oxygen in the lower layers is reduced to very low levels if there is a large oxygen demand by decomposable organic matter that settles to the lake bottom. Localized low oxygen conditions occur in streams when water is released from large lakes at the bottom of the dam. Such water is low in dissolved oxygen, and flows some distance downstream before it is re-aerated to restore normal levels of oxygen.

Measurement of oxygen in Georgia streams began, along with most other water quality monitoring, in the 1960s. Since that time, oxygen in most streams has increased or remained the same. Figure 4.4 shows three examples of trends since the 1970s. Dissolved oxygen in the Chattahoochee at Fairburn and the South River at Bouldercrest Road in southeast Atlanta has more than doubled. Improvement in the Chattahoochee River is even greater than shown in Figure 4.4. In a survey conducted in 1963 in the stretch 20–40 miles south of Atlanta, it was found that oxygen was consistently below 4.0 milligrams per liter.[21] Minimum oxygen concentrations measured at Fairburn in three consecutive years, 1969 through 1971 were below 1.0 milligram per liter. However, since 1990 minimum oxygen concentrations at this location have not dropped below 7.0 milligrams per liter and the averages shown in Figure 4.4 have been around 9.0 milligrams per liter.

Oxygen in the lower Savannah River was as low as any large waterway in Georgia in the 1960s. For about a 10-mile stretch near the city of Savannah, dissolved oxygen for the month of August averaged from 1.8 to 3.7 milligrams per liter during 1967, 1968, and 1969.[22] Further up the river near Clyo, above the influence of the city and most of the industries, the yearly average of dissolved oxygen was never below 6.5 milligrams per liter, but it decreased from about 9.0 milligrams per liter in the late 1960s to about 7.5 milligrams per liter in the mid-1980s and has changed little since then. The decrease near Clyo in the 1970s and 1980s was apparently not due to increased BOD, since BOD at this location has not changed appreciably since the 1960s. Oxygen may have dropped

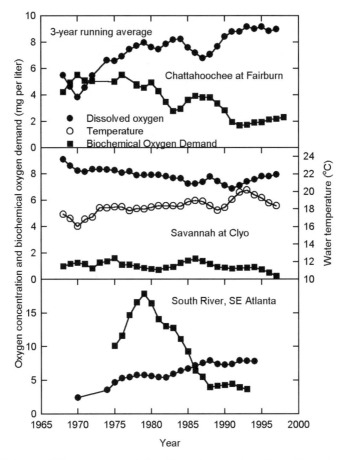

Figure 4.4. Changes with time in oxygen concentration and biochemical oxygen demand in three Georgia rivers. Data are from records of the US Geological Survey.

because of the 3 to 4 °C increase in river temperature during this period (Figure 4.4). Of the fifteen river sites I examined data for, only one had annual average oxygen concentrations below 7.0 milligrams per liter in the 1990s. That location was on the Satilla River at Atkinson, where oxygen appears to have decreased rather steadily since the 1970s, and the average was below 7.0 from 1990 to 1995. The decrease in oxygen does not correspond to changes in water temperature, because temperature of the Satilla has not changed since the 1970s.

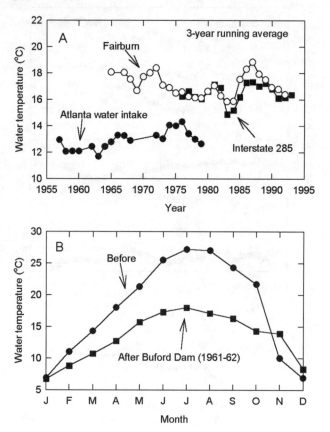

Figure 4.5. Changes in temperature of the Chattahoochee river in Atlanta above and below two electric power generating plants (A) and before and after construction of Lake Lanier (B). Data for (A) from US Geological Survey records, and for (B) from US Department of the Interior (1966).

Increases in temperature cause problems for maintaining high oxygen levels in some Georgia localities, mainly downstream from electric power generating plants. Electric power generators use river water for cooling of reactors and the cooling of the equipment heats the water. The heating is illustrated in Figure 4.5A by the difference in average yearly temperature of the Chattahoochee at the Atlanta Water Works intake and at Interstate 285 and Fairburn. The Atlanta water intake is 5.3 miles upstream from Interstate 285 and 18 miles upstream from Fairburn. The Atkinson and McDonough steam power plants operated by Georgia Power Company are about 2 miles upstream from Interstate 285. So, the river temperature rises 2 to 4 $^{\circ}$C (about 3.5 to 7.0 $^{\circ}$F.) between the Atlanta water intake and Interstate 285, and remains at that higher temperature from Interstate 285 to Fairburn. This increase in temperature caused by the power plants reduces the oxygen concentration of the Chattahoochee, but if oxygen is not too low, higher temperature may increase the breakdown of waste it receives from the metropolitan area. In this case of temperature pollution, the negative effects have been partially offset by releases of cooler water from Lake Lanier. As shown in Figure 4.5B, the river temperature at Atlanta was reduced after completion of Lake Lanier in 1957. The river was cooler by almost 10 $^{\circ}$C in mid-summer of 1961 and 1962 compared to years before the dam was built, but mid-winter temperatures were hardly affected.

Because of the complicated conditions that determine oxygen concentration in streams, oxygen cannot be taken as the sole indicator of water quality. Some streams can be quite polluted and still have favorable oxygen concentrations. Peachtree Creek in Atlanta would be considered polluted based on its yearly average fecal coliform counts of between 10,000 and 20,000 per 100 milliliters (Figure 4.3). However, it has consistently maintained annual oxygen concentrations greater than 7.0 milligrams per liter and in most cases above 8.0 milligrams per liter since the 1970s. Even high BOD does not necessarily lower oxygen. Peachtree Creek had oxygen levels above 8.0 milligrams per liter in the years 1976 and 1977 when BOD levels were 6.2 milligrams per liter, higher than in the Chattahoochee at Fairburn (Figure 4.4). Note also that, in the South River, BOD rose from about 10 to 17 milligrams per liter during the late 1970s when oxygen concentration was actually increasing. The lack of agreement between low oxygen and high BOD concentrations arises partially from the fact that oxygen is depleted in the stream where organic matter is decomposed, which may be several miles downstream from where it is dumped or measured.

It may be noticed in Figure 4.4 that BOD was never very high in the Savannah River near Clyo, but the amount of organic matter dumped in the Savannah and other Georgia streams has declined over the years. The Georgia Department of Natural Resources (1976) estimated that BOD loading to the Savannah below Clyo from major sources, such as industry and municipalities, decreased from 208,000 pounds per day in 1965 to 23,000 pounds per day in 1975. Release of BOD from Atlanta waste-water treatment plants decreased from 20 metric tons per day in 1976 to 12 metric tons per day in 1985, even though population increased greatly and waste water volume increased by 60 percent.[23]

Reducing Nitrogen and Phosphorus

Fecal bacteria are not the only pollutants resulting from release of sewage into streams, and even treatment of sewage does not remove all of the pollutants. Two of the elements in sewage that are important to the health of streams receiving the sewage are nitrogen (N) and phosphorus (P). These elements are vital in small amounts to all life forms. They are consumed normally in foods, but in large amounts, the nitrate form (NO_3) of nitrogen is toxic. The most notable health problem of too much nitrate is methemoglobinemia in infants or "blue baby" syndrome. The actual cause of the sickness is nitrite (NO_2), which is a similar molecule that is formed from nitrate in the body by chemical reduction. Nitrite is normally present in the body in very small amounts, and in larger amounts interferes with the ability of the blood to carry oxygen, resulting in sickness and the associated abnormal color of the blood. Babies are much more sensitive to high levels of nitrate than adults. The upper safe limit of nitrate in drinking water set by the US Enviromental Protection Agency (USEPA) is 10 milligrams of N per liter.

In a recent publication, Avery (1999) questioned the toxicity of nitrates in drinking water and hypothesized that gastrointestinal infection and inflammation that caused overproduction of nitric oxide likely caused the cases of methemoglobinemia in infants attributed to drinking water. If the hypothesis is correct, according to Avery the USEPA standard for nitrates in drinking water should be reevaluated. There have been many surveys of drinking water quality based on the dangers of nitrate, although there have been only two cases (not deaths) of methemoglobinemia linked to drinking water in the United States since the 1960s,[24] neither of them in Georgia.

Although phosphorus is not a health hazard for humans and animals, large amounts of both phosphorus and nitrogen are deleterious to the health of streams and lakes. The problem is mainly one of too much stimulation of growth of algae and other aquatic plants. When algae and other plants die, settle to the bottom, and decompose, they use oxygen from the water and deplete the oxygen supply required by fish and other organisms in the streams and lakes. The process of enrichment of lakes with minerals, the growth of algae that results, and depletion of oxygen supply is called eutrophication. It is a natural process that follows creation of lakes, because of the nutrients delivered to the lakes by streams entering them. In the natural process of eutrophication, the nutrients come from weathering of rocks, leaching of soils, and even from rainfall, as we shall see later in the chapter on air quality.

The water pollution problem of nitrogen and phosphorus is that they accelerate the eutrophication of streams and lakes. This acceleration is sometimes referred to as "cultural eutrophication." High levels of these two elements in streams can come from several sources; sewage (treated or not), agricultural or urban areas fertilized with these nutrients, wastes from livestock or other animals, and effluent from some processing plants. In the past, contamination of streams by nitrogen and phosphorus generally was much worse than now, although the trends differ a great deal for the two elements and for different locations in the state.

A Kellogg Interdisciplinary Task Force on Physical Resources in Georgia[25] concluded nonpoint source pollution, especially from agriculture, to be a growing problem for Georgia waterways. Nonpoint source pollution is that which washes off the land rather than coming from sewers and other well-defined points. It was stated in the Kellogg Report that "…nonpoint source pollution, as indexed by nitrates, has increased in all 5 Georgia rivers that have been monitored for 25 years or more." The example for this statement in the report was for the Savannah River during the period 1975 to 1985. The statement in the report was only partially true, and the example of the Savannah River was not. The data in Figure 4.6 show the nitrate (plus nitrite which makes up a very small portion of the total) trends for four locations on rivers in Georgia. In the Savannah River near Clyo, nitrate changed little from 1970 to 1998; but the change was downward (Table 4.1). There also appears to be little increase in nitrate in the Flint River at Montezuma just below the Fall Line and at most other river locations over the same period. One of the most striking examples of decrease in nitrogen pollution is in the South River at Bouldercrest Road in southeast Atlanta (Figure 4.6). The average nitrate concentration in that river was 3 milligrams per liter in 1977, but dropped to between 0.5 and

1.0 milligram per liter after the late 1980s. This decrease in nitrate concentration resulted from better sewage treatment in the late 1970s and 1980s, and transfer of treated wastewater to the Chattahoochee rather than the South River.

Analysis of the change in nitrate in several of the larger or most studied streams shows no consistent trends from the 1970s to the 1990s. Of the thirty-nine tests for trends listed in Table 4.1, eighteen showed increasing nitrate, seven had decreasing nitrate and fourteen showed no change. Of the eighteen with increasing nitrate, seven were in the Atlanta metropolitan area or in the Chattahoochee River south of Atlanta. Increases were small at most locations where they occurred, with over half showing 0.1 milligram per liter change or less in ten years. Frick et al (1996) did not test trends over the entire period from the early 1970s to 1990 (Table 4.1), but in most cases where I examined records for the same location the results

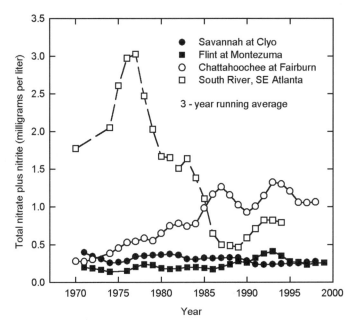

Figure 4.6. Trends in nitrate in four Georgia rivers (includes the nitrite form which makes up a small proportion). Data are from the US Geological Survey records.

Table 4.1 Changes in nitrate (milligrams per kilogram) in Georgia streams.

Location	Period tested	Change in nitrate mg/liter/yr	Median conc. mg/liter	Source[1]
Chattooga River near Clayton	1968-83	none	0.04	1
Conasauga River at Tilton	1971-99	none	0.42	3
Etowah River at Canton	1970-96	+0.010	0.21	3
Coosa River near Rome	1970-96	+0.007	0.33	3
North Oconee River at Athens	1975-83	none	0.30	1
Oconee River at Milledgeville	1970-99	none	0.19	3
Ocmulgee River near Macon	1974-99	none	0.45	3
Falling Creek near Juliette	1971-83	+0.002	0.05	1
Chattahoochee River near Cornelia	1980-90	+0.013	0.20	2
Chat. River (Gwinnett Water Intake)	1980-90	+0.003	0.22	2
Chat. River (Dekalb Water Intake)	1980-90	+0.005	0.21	2
Chat. River (Cobb Water Intake)	1980-90	none	0.30	2
Chat. River (Atlanta Water Intake)	1980-90	+0.003	0.32	2
Chattahoochee River at I-285 (W)	1975-95	+0.034	0.47	3
Chattahoochee near Fairburn	1970-99	+0.036	0.88	3
Chattahoochee at Whitesburg	1980-90	+0.090	0.67	2
Chattahoochee at West Point	1970-95	none	0.35	3
Chat. River (Columbus W. Intake)	1980-90	none	0.35	2
Peachtree Creek at Atlanta	1980-90	none	0.51	2
Peachtree Creek at Atlanta	1970-98	+0.008	0.53	3
Sweetwater Creek at Austell	1968-83	−0.008	0.27	1
South River in S.E. Atlanta	1970-93	−0.101	1.03	3
Flint River near Jonesboro	1980-90	−0.139	0.80	2
Flint River near Fayetteville	1980-90	−0.116	0.80	2
Flint River near Inman	1980-90	−0.097	0.72	2
Flint River above Griffin	1980-90	−0.077	0.60	2
Flint River at Montezuma	1971-99	+0.005	0.22	3
Flint River near Newton	1971-95	none	0.37	3
Flint River near Albany	1980-90	none	0.27	2
Flint River near Putney	1980-90	+0.016	0.37	2
Altamaha River near Everett	1970-99	+0.004	0.24	3
Ogeechee River near Eden	1971-97	none	0.09	3
Ogeechee River near Eden	1971-91	+0.01	0.10	4
Savannah River near Clyo	1971-98	−0.003	0.30	3
Ochlocknee River near Thomasville	1971-93	none	0.72	3
Withlacoochee R. near Valdosta	1974-93	none	0.17	3
Swannee River at Fargo	1971-91	+<0.01	0.00[2]	4
Satilla River at Atkinson	1971-98	+0.003	0.08	3
Canoochee River near Claxton	1971-91	+<0.01	0.00[2]	4

[1] References listed below. [2] Median values less than 0.1 mg per liter.
1 - Buell and Grams, 1985. 2- Frick et al., 1996. 3 - My analyses of U.S. Geological Survey data. 4 - Ham and Hatzell, 1996.

were similar for the longer period. When trends were similar for the same location in two different studies, only the results for the longer period are shown in Table 4.1. In two cases, analyses of data for the same station gave different results. In the Ogeechee River in Southeast Georgia near Eden, my analysis of the data showed no trend, whereas Ham and Hatzell (1996) found an increase in nitrate. On the other hand, for Peachtree Creek in Atlanta I found an increase in nitrate, whereas Frick et al found no trend for the shorter period, 1980 to 1990. Such differences for the same locations mean that the increases are small enough to be questionable.

Not only are the changes in nitrate in most Georgia streams slight or non-existent, but actual concentrations are quite low. Median nitrate concentrations in Table 4.1 are not averages, but are levels for samples that rank halfway between the highest and lowest from that location. The median concentrations in Table 4.1 are all less than 1.0 milligram per liter, except for the South River that was so badly polluted in the 1970s. The most recently determined nitrate concentrations in the South River are between 0.5 and 1.0 milligrams per liter (Figure 4.6). For those locations with increasing nitrate concentrations, the median concentrations range from less than 0.1 to 0.88 milligrams per liter. Among these locations the highest median and current concentrations are for the Chattahoochee and Flint Rivers south of Atlanta.

Although treatment of sewage is not very effective in removing nitrogen, in the last thirty years there has been more improvement than the changes in nitrate levels show. The ammonium form of nitrogen is much more harmful to fish and other aquatic organisms than nitrate. In modern treatment of sewage, much of the ammonium form of nitrogen is converted to nitrate. In the early 1970s, the concentration of ammonium nitrogen was higher than nitrate in some streams. In the South River at Bouldercrest Road, for example, this form was twice as high as nitrate in 1970. By the late 1980s, however, it was less than 1 milligram per liter and similar to the level of nitrate. A similar trend occurred in the Chattahoochee near Fairburn. Whereas ammonium nitrogen was higher than nitrate in the early 1970s (Figure 4.6), by the mid-1990s ammonium nitrogen was only about 0.25 milligram per liter.

The conclusion and example of the Kellogg task force would have been much more correct if it had referred to the Chattahoochee River below Atlanta (near Fairburn) where the increase in nitrate has been rather steady. The largest increases in nitrate from 1980 to 1990 in the Apalachicola-Chattahoochee-Flint river basin was in the Chattahoochee from Atlanta to Columbus.[26] However, it appears that agriculture is not the major source of nitrates in the Chattahoochee River and most other Georgia streams. Frick et al (1996) concluded that the highest annual yields (input to the river) for nitrogen in the Apalachicola-Chattahoochee-Flint River basin were in the Chattahoochee River south of Atlanta. Atlanta is the largest source of nitrogen for that large river basin, although agriculture and forestry account for 88 percent of the area.

Ham and Hatzell (1996) studied nitrate in several streams in the Coastal Plain and found increasing nitrate since 1971 (Table 4.1), although the median concentrations were low (all less than 1.0 milligram per liter). The source of increasing nitrate in these streams was apparently urban, however, not agricultural. They were surprised at the finding, stating that "The lack of association between high nutrient concentrations and the agriculture category was not expected since the removal of nutrients in runoff from agricultural production has been documented." In 1989, the Georgia Environmental Protection Division ranked Gum Branch Creek in Crisp County among streams given first priority for management and demonstration projects based on intense use of fertilizers and pesticides.[27] However, Cofer et al (1991) found that the urban area, which accounted for only 7 percent of the Gum Branch Creek watershed contributed 90 percent of the nitrogen and 98 percent of the phosphorus put into the stream. Another indication that the increases in nitrate in streams are not likely the result of increased agricultural fertilizer use is that nitrogen fertilizer use in Georgia has decreased by over 50 percent since the mid-1970s (Figure 4.7). In addition, the most significant increases in nitrate concentrations have been in Piedmont locations, where agricultural use of nitrogen has likely decreased even more, because row crop agriculture has shifted to the Coastal Plain.

Another source of nitrate in streams mentioned by the Kellogg Task Force is waste from large concentrations of animal and poultry operations. There has been an increase in numbers and concentration of farm animals with the conversion from row crops to more diversified agriculture. The poultry industry has grown, especially in north Georgia, to the largest agricultural enterprise in the state. This growth in numbers and concentration of animals in large operations

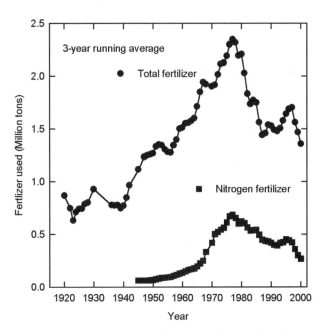

Figure 4.7. Trends in use of total and nitrogen fertilizer in Georgia. Data are from the US Bureau of the Census (1920–2000), and the Georgia Agricultural Statistics Service (1950–2000).

has led to speculation by several study groups that agriculture is a nonpoint source of pollution of waterways. However, several studies have shown no good evidence that nutrients from animal confinement has caused widespread decreases in stream quality. The Georgia Nonpoint Source Management Plan of 1989 concluded that "Nonpoint source pollution from nutrients and pesticides is not a significant problem in Georgia at this time." Smith and Sellers (1995) studied four streams in the Lake Lanier drainage basin and concluded as follows. "The data collected during this study do not provide strong evidence that the high concentrations of poultry houses and the associated land application of litter are contributing to a degradation of the streams that were monitored. Almost all of the significant differences were associated with Mud Creek, and these differences are likely due to the discharge of municipal waste treatment plant effluent into the stream." Shellenberger et al (1996) concluded that, "Sampling of streams at low flow stages in three counties with the largest poultry production or dairy livestock production [Barrow, Lumpkin, and Morgan counties] failed to show any significant impact from such agricultural activities on nitrate levels." The situation appears to be that there is little effect of these animal concentrations on water quality, except in the immediate vicinity of the animal pens and houses.

Nitrogen from polluting sources may also enter natural underground reservoirs, called aquifers. Much concern has been expressed about contamination of underground reservoirs because of increased use of fertilizers and increases in animal numbers, especially chickens. However, there is little evidence of increased contamination in spite of the large numbers of wells analyzed. One of the reasons for the widespread speculation rather than firm conclusions about nitrate pollution of underground water is that no systematic study has been made of these underground reservoirs to determine if nitrate has changed over time. The US Geological Survey has sampled several hundred wells over the state, but the same wells have not been sampled regularly for long periods of time. To see if nitrate contamination has increased in Georgia wells, I have summarized the nitrate analyses of the US Geological Survey from the 1940s through the 1970s (the US Geological Survey terminated monitoring in the early 1980s) and compared them with two large surveys in the 1990s (Table 4.2). Wells less than 100 feet deep were separated in the summary because shallow wells are the ones most likely to be contaminated with nitrate. The number of wells less than 100 feet deep analyzed in each decade from 1940 through the 1970s is small and it is uncertain how accurately they represent the whole state. However, there is no indication of increased pollution of underground water by nitrate in the shallow

Table 4.2 Nitrate concentration (milligrams per liter) in wells in Georgia from the 1940s to the 1990s.

Period	No. wells	Ave. conc. (mg/L)	% exceed. 10 mg/L	Source
Shallow wells (0 to 100 ft.)				
1940s	28	1.17	0	U.S. Geological Survey
1950s	22	2.32	4.5	"
1960s	92	1.66	3.3	"
1970s	20	0.45	0	"
1989-93	1,120	1.16	3.8	Tyson et al., 1995
Depth not given (less than 250 ft.)				
1990-95	5,072	0.99	1.2	Shellenberger et al., 1996.
Deep wells (deeper than 100 ft.)[1]				
1940s	77	0.31	0	US Geological Survey
1950s	208	0.44	0	"
1960s	689	0.31	0.1	"
1970s	276	0.33	0	"
1989-93	2,299	0.48	0.9	Tyson et al., 1995

[1] For U.S. Geological Survey wells, 100 to 1,000 ft.

or deep wells in Table 4.2. The average nitrate for 1,120 shallow wells sampled from 1989 to 1993, was 1.16 milligrams per liter. This is lower than the average of 1.51 milligrams per liter for the total of 162 shallow wells analyzed prior to the 1980s. Less than one percent of wells deeper than 100 feet had nitrate concentrations higher than the level considered safe for drinking (10 milligrams per liter). Three to four percent of shallow wells had higher than 10 milligrams per liter of nitrate in the 1950s, 1960s, and 1990s.

Shellenberger et al (1996) conducted a large survey of wells in 146 counties from 1990 to 1995. They excluded thirteen counties along the coast and southeastern Georgia "where there is little agriculture and where domestic drinking water wells are typically drilled to the deeply confined Floridan aquifer." They also excluded the urbanized core of metropolitan Atlanta because well usage is rare in the area. The depth of individual wells was not stated in the Shellenberger et al report, but all wells were less than 250 feet deep. They found more than 10 milligrams per liter of nitrate in only 1.22 percent of the samples. When nitrate concentrations were higher than 5 milligrams per liter, a retesting of the well was done after several months and the area was investigated for possible sources of contamination. Seventy-four percent of 210 wells retested had nitrate concentrations lower than the first test. Wells high in nitrate were found to be in close proximity of septic tanks, animal enclosures, and heavily fertilized areas (especially those covered with chicken litter). Although the percentage of wells with high nitrate is low in the surveys of both the University of Georgia Cooperative Extension Service[28] and the Georgia Geological Survey[29] they may be higher than the actual percentage of wells above the drinking water standard. The well-testing service was advertised by both of the testing agencies and many of the well owners who sent samples may have done so because they suspected contamination. So the percentage of wells with evidence of contamination may be higher than would be the case if wells were selected at random.

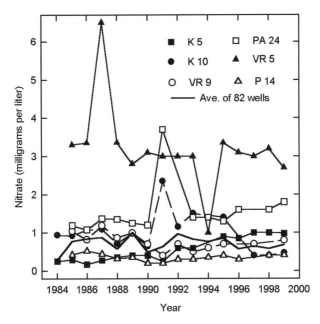

Figure 4.8. Trends in nitrate in selected wells in Georgia. Wells are located in the counties of Richmond (K5), Peach (K10), Polk (VR9), Chatooga (VR5), Decatur (PA24), and Upson (P14). The heavier solid line is the average of eighty-two wells during the sixteen-year period. Data from the Georgia Geological Survey (1985–2001).

The change of nitrate in wells over time is shown more accurately if the same wells are tested regularly for a long time. The Georgia Geological Survey (1985–2001) started a systematic study of wells in Georgia in 1984. The number of wells was small compared to some other studies, some were substituted for by other wells in the same area in later years, and the time period is too short to establish long-term trends. In spite of these shortcomings, the data are valuable because they indicate no widespread pattern of increased nitrate contamination of wells. In Figure 4.8, six representative wells are shown. Wells were chosen to represent different areas of the state and to show the range of nitrate found. In selecting individual wells for Figure 4.8, those where substitutions were made during the sixteen-year period were omitted (see exception below). It is clear that those wells that were high in nitrate in 1984 were still high in 1999 and those that were low in 1984 were still low. There were slight up or down trends, but changes were less than 0.5 milligram per liter, except for well K5 in Richmond County in which nitrate has increased from 0.25 to 1.0 milligram per liter in the sixteen-year period. The heavy line in Figure 4.8 represents the average of eighty-two wells in the Georgia Geologic Survey's study, most of which were sampled less often than the individual wells shown in the figure. It is also clear from the overall average of these eighty-two wells that there was no appreciable increase in nitrate since 1984.

Another point illustrated by Figure 4.8 is that contamination of wells is localized rather than widespread. Occasionally much higher or much lower nitrate concentrations are measured than are characteristic of a well, as shown by the peak for well VR5 in 1987 and the drop in 1994. If these peaks and valleys reflect real changes in nitrate contamination of the well, then the effects must be restricted to a localized area. Otherwise, the changes would not be expected to be so short-lived. In addition, when well K10 (Fort Valley well #1) was substituted in 1997 by Fort Valley well #5, nitrate dropped from 1.4 to 0.4 milligram per liter since the previous sampling (1995). It remained at the same level in 1998 and 1999. While the drop might represent a decrease in nitrate in Fort Valley's water supply that coincided with the change in test wells, it is more likely that two wells in the same town have different nitrate levels.

This localized nitrate contamination of well water is seen in areas where sources of nitrogen are rather obvious. Wells, especially shallow ones, located close to large amounts of animal wastes have higher nitrate concentrations. In the Little River/Rooty Creek watershed in east-central Georgia where there is a high concentration of dairies, a survey of 284 wells showed a higher incidence (12 percent) of nitrate levels above the drinking water standard.[30] Visual inspection showed that the wells with high nitrate levels were close to cattle loafing areas, septic drain fields, or were influenced by surface water runoff. Bush et al (1997) also noted that follow-up investigation of heavily contaminated wells in Georgia usually revealed a septic system septic system or animal wastes near the well. So there are local conditions that cause high nitrate concentrations in well water, but widespread or increasing contamination of underground water by runoff or leaching from agricultural fields or other nonpoint sources is still a matter of speculation.

In some analyses, there are also differences in nitrate concentrations in wells located in different sections of the state. In the coastal flatwoods of Southeast Georgia and in the northwestern section of the state Tyson et al (1995) found shallow wells (less than 100 feet deep) had an average of 3.3 milligrams per liter of nitrate compared to the 1.16 milligrams per liter for the state average. However, in the survey of Shellenberger et al (1996) there were no apparent differences in the various sections of the state. Contamination of wells can result if they are in or near fertilized fields, but even that is not likely to cause high concentrations. Nitrate concentrations in half of sixty-one shallow (less than 80 feet deep) monitoring wells adjacent to farm fields in the southwestern part of the state were below 3.0 milligrams per liter, a concentration considered to be the breaking point between natural conditions and human influence.[31] Fifteen percent of the wells had concentrations above the drinking water limit. This 15 percent is higher than in most surveys and probably resulted from the location adjacent to fertilized fields, the sandy nature of the soil, and the shallow depth.

There have been numerous references in recent years to the possibility or probability of widespread ground water contamination by agricultural fertilizer and animal manure use. There is, however, little evidence that points to such contamination. An analysis of the survey data of Shellenberger et al (1996) indicates that those counties with the most intensive row-crop cultivation do not have more nitrate in wells. The percentage of land in harvested crops in 1992[32] is plotted in Figure 4.9 against the percentage of wells surveyed from 1991–1995 that had more than 3.0 milligrams per liter of nitrate. There were twenty-one counties with 20 percent or more of their land area in harvested crops and all except Bulloch County were in the southwest quadrant of the state, the section identified in the 1970s and 1980s as the most likely to suffer from nonpoint source pollution.[33] None of the twenty-one counties had more than 12 percent of wells above the 3.0 milligrams per liter nitrate level. On the other hand, the twenty-six counties that had 12 percent or more of their wells above 3.0 milligrams per liter were scattered rather evenly across the state. In a separate analysis, thirty counties rated by the Georgia Nonpoint Source Management Plan (1989) as having the most "pollution potential" from agricultural nonpoint sources did not have significantly more wells over 3.0 milligrams per liter of nitrate (8.2 percent) than the state average (7.1 percent), and none were among the top ten counties for contamination with nitrate. Dooly County, which was ranked as having the

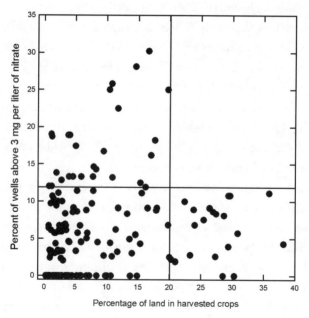

Figure 4.9. Nitrate in wells from 1991–1996 related to the percentage of land in harvested crops in 1992 for 144 Georgia counties. Nitrate data from Shellenberger et al. (1996); harvested cropland from the US Bureau of the Census (1920–2000).

greatest potential for degraded water quality from agricultural non-point sources,[34] had only two out of forty-six wells (4.3 percent) above the 3 milligrams per liter level and none above the USEPA standard for drinking water. While these analyses do not rule out field crop and animal agriculture as contributors to nitrate contamination of wells, they minimize agriculture as the source and indicate that other sources are likely to contribute more.

The main concern of contamination of waterways with phosphorus is the eutrophication and detrimental effects on aquatic organisms. The area of most contamination in Georgia is downstream of the Atlanta metropolitan area (see Figure 4.2B). Amendments to the Georgia Water Quality Control Act directed major wastewater treatment plants to reduce the average phosphorus concentration in effluent into the Chattahoochee River to 0.75 milligram per liter or less by 1 January 1992.[35] By 1993, all treatment plants were in compliance, except three owned by the City of Atlanta. The city negotiated an extension until 4 July 1996, in exchange for agreeing to meet a more restrictive limit of 0.64 milligram per liter average phosphorus concentration in the effluent. In spite of the noncompliance of some of the wastewater treatment plants in the Atlanta area and the frequent spills in the sewage treatment system, phosphorus contamination of the Chattahoochee has decreased. DeVivo et al (1995) calculated that the phosphorus discharged to the Chattahoochee in the metropolitan Atlanta area decreased by about 83 percent from 1988 to 1993 because of legislation restricting the use of phosphorus detergents, and the improved removal of phosphorus by water treatment plants.

The USEPA has recommended 0.10 milligram per liter of phosphorus as a maximum concentration to control eutrophication in flowing waters and 0.05 milligram per liter to control eutrophication where streams enter a lake or reservoir. It is interesting that although the USEPA recommends 0.10 milligram per liter as the maximum concentration for flowing streams, the US Geological

Survey considers it a background (natural) level.[36] More than 40 percent of the samples taken in a 1994–1995 study of the Chattahoochee River basin by Peters et al (1997) exceeded the 0.10 milligram per liter recommendation and 75 percent of the samples exceeded the level for waters entering a lake or reservoir. Likewise some locations I examined in US Geological Survey records on other rivers exceeded one or both of those recommendations when averaged over the latest five years of sampling (Table 4.3). High concentrations of phosphorus were not restricted to any section of the state, the two highest being from the Conasauga River in northwest Georgia and the Ochlockonee River near Thomasville in the southwest. The Flint River in southwest Georgia and the Ogeechee and Altamaha Rivers in southeast Georgia were among the lowest in phosphorus concentrations. These patterns of distribution of high and low phosphorus concentrations point to local rather than regional sources of phosphorus contamination. Frick et al (1996) found the highest mean annual yields (concentration times volume of water) for phosphorus in the Apalachicola-Chattahoochee-Flint River basin south of Atlanta.

Although phosphorus levels were above USEPA Environmental Protection Agency stream guidelines in eight of the eighteen places listed in Table 4.3, the latest levels were substantially lower than the maximum levels from earlier years. In several streams there have been substantial increases and decreases in phosphorus over the years, as in the Chattahoochee River near Fairburn south of Atlanta and in the Conasauga River near Tilton (Figure 4.10). Phosphorus in the Chattahoochee near Fairburn was nearly constant at 0.5 milligram per liter from the time measurements began in 1971 to about 1982. There was a subsequent increase, however, to 1988 that was likely caused, in part, by the diversion of treated sewage from the South River to the Chattahoochee mentioned earlier for nitrate. Since 1988, phosphorus has decreased in the Chattahoochee near Fairburn from 0.88 milligram per liter to 0.123 milligram per liter in 19987. The most remarkable reduction of phosphorus I found in records of water analysis was in the South River. When records began in 1970, phosphorus was over 4.0 milligrams per liter. It declined by 95.3 percent from the mid-1970s to the early 1990s (Table 4.3). In some other rivers, like the Etowah River at Canton north of Atlanta, phosphorus was never very high (Figure 4.10). A recent study of water quality in Lake Allatoona indicated, without presenting data, that phosphorus was higher in the Etowah in the

Table 4.3 Changes in phosphorus (micrograms per liter) in Georgia streams from the highest consecutive-5-year average to average of the last five years of sampling. Data from the U.S. Geological Survey records.

Location	HIGHEST 5 YEARS Years	Phosphorus mg per liter	LAST 5 YEARS Years	Phosphorus mg per liter	% change
Conasauga at Tilton	1983-87	1.128	95-99	0.354	-68.6
Etowah at Canton	1989-93	0.087	92-96	0.072	-17.2
Coosa near Rome	1970-74	0.193	91-96	0.151	-21.8
Peachtree Creek	1974-78	0.185	94-98	0.133	-28.1
Chattahoochee at I-285	1985-89	0.471	91-95	0.151	-67.9
Chat. near Fairburn	1985-89	0.701	94-98	0.124	-82.3
Chat. at West Point	1971-75	0.153	91-95	0.071	-53.6
Flint at Montezuma	1988-92	0.097	95-99	0.039	-59.8
Flint at Newton	1986-90	0.086	91-95	0.039	-54.6
South River - S. E. Atlanta	1974-78	2.879	89-93	0.135	-95.3
Oconee at Milledgeville	1989-93	0.089	91-99	0.066	-25.8
Savannah near Clyo	1985-89	0.121	94-98	0.100	-17.3
Ocmulgee above Macon	1975-79	0.103	95-99	0.046	-55.3
Ogeechee near Eden	1986-90	0.077	91-97	0.059	-23.4
Altamaha at Everett	1987-91	0.091	92-99	0.070	-23.1
Satilla at Atkinson	1985-89	0.130	90-98	0.088	-32.3
Ochlocknee nr. Thomasville	1977-81	1.109	89-93	0.527	-52.5
Withlacoochee nr. Valdosta	1977-81	0.377	88-92	0.212	-43.8

mid-1990s than earlier.[37] However, no such trend appears in the US Geological Survey data plotted in Figure 4.10.

Changes in phosphorus in the streams over the last half of the twentieth century have been variable, but mostly downward. Because changes in phosphorus have not been steady over the years, I have analyzed the changes in several major streams by comparing the average of the most recent five years of sampling in the US Geological Survey records with the highest consecutive-five-year average. In all of the locations listed in Table 4.3, there was a decrease in phosphorus concentration from the highest to the most recent five years, and in half of the eighteen locations the decrease was more than 50 percent. In some cases the highest concentrations were in the early 1970s, when testing began; in others highest concentrations were observed in the late 1980s and early 1990s. There were also cases where phosphorus concentrations were much lower in years before the maximum occurred, as in the Chattahoochee

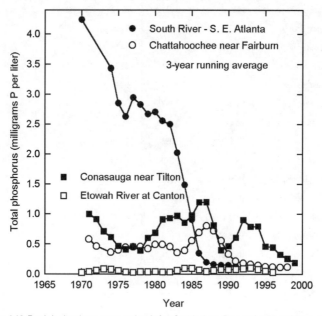

Figure 4.10. Trends in phosphorus concentrations in four Georgia rivers. Data are from the US Geological Survey records.

River near Fairburn and the Conasauga River (Figure 4.10). Decreases in phosphorus in Georgia's streams may result from many changes, but the two that have likely made the most impact are the banning of phosphorus detergents and better wastewater treatment. This is clearly illustrated by the drop in phosphorus concentrations at locations downstream of Atlanta during a period when the population of the metropolitan area increased by about 70 percent.

Removal of Industrial Pollutants and Pesticides

Industrial pollutants are widespread, especially those that are put into the atmosphere and are scattered by the wind. However, the most serious problems tend to be located fairly near a major pollutant source. As one would expect, most of the pollution of streams with industrial contaminants was localized downstream from industrial centers. Industrial pollutants were released into streams of Georgia in large quantities during the twentieth century. There was local concern in many cases early in the century about industrial pollution of streams, but perception of the dangers did not become state wide or national until mid-century. Testing for industrial pollutants in streams began only in the 1970s, as was the case for most pollutants.

Industrial pollutants include heavy metals, such as mercury, zinc, chromium, and lead, and polychlorinated biphenyl compounds referred to as PCBs. One of the main concerns for pollution by these substances is that they persist a long time in streams and lakes. A substantial amount of lead in the past apparently came from the use of leaded gasoline, and it was deposited on the land by rainfall and into the streams by washing off of city streets, parking lots, and highways. Mercury is released into the environment from natural sources such as volcanoes, ocean waters, and the earth's minerals. It is mined as the mineral cinnabar in several countries. But the sources of main concern are man-made. Mercury pollution comes from manufacture and disposal of many products such as batteries, light bulbs, thermometers, pesticides, and paint. It is released from the burning of fossil fuels, lead smelters, and the production of chlorine. PCBs were released from industrial processes and especially from capacitors used in electric power transmission. Mercury is a poisonous heavy metal that causes damage to the central nervous system and to the kidneys. One of the odd ironies of mercury use is that an element now taken so seriously as an environmental pollutant was, for more than a thousand years, valued as a medicine. Its use as a cure for syphilis led to symptoms of mercury poisoning that in past centuries were confused with symptoms of the disease itself. D'Itri and D'Itri (1977), state that

> Calomel [mercurous chloride] was…used to treat syphilis, as a laxative, and briefly as a diuretic before being discontinued in the nineteenth century because of its negative side effects. In the United States mercurous chloride became an almost universal panacea. Medical textbooks of the last century list this all-purpose drug as a primary cure for such diverse illnesses as fever, diarrhea, heart disease, worms, rheumatism, and eye diseases.

These authors were wrong about the discontinuance of use of calomel (mercurous chloride) in the nineteenth century. Many children of the first half of the twentieth century in Georgia remember taking calomel at least once a year to "clean out the system."

Mercurochrome and Merthiolate were popular disinfectants during much of the twentieth century, although they were found after extensive testing to merely delay multiplication of bacteria rather than destroying them. These disinfectants contained up to 25 percent mercury. In addition to the extensive use of mercury in fillings of tooth cavities, dental practitioners in the past used mercury compounds to sterilize instruments. In a survey of dental schools and dentists in 1937, 63 percent favored use of mercurial disinfectants for instruments that would be damaged by boiling.[38] Thus, although mercury is certainly toxic, it is not acutely toxic.

Otherwise its medicinal use would have been discontinued centuries ago.

Mercury is found in sediments of streams below metropolitan areas and below certain industries. One of the main sources of mercury in streams in the past was chlor-alkali plants that used it in the manufacture of chlorine and caustic soda. In addition, large amounts were used as a fungus or slime retardant in the manufacture of pulp and paper. It was also used extensively in mining, especially to amalgamate silver and gold in the recovery process. Mercury residues still remain in the gold-mining area around Dahlonega, with contamination of sediment in or near affected streams in the range of 0.02 to 12.0 milligram per kilogram, however, it decreases rapidly downstream to concentrations that are only slightly above background levels.[39]

Some fish accumulate high levels of mercury in their bodies, especially if they prey on other fish. If the concentrations rise to high enough levels, the health of humans and wildlife consuming the fish are endangered. The US Food and Drug Administration (FDA) set a maximum safe level of total mercury in fish for human consumption of 0.5 milligram per kilogram in 1969. Establishment of "action levels" in fish came as a result of high levels of mercury found in fish, especially tuna and swordfish, and a few incidents of mercury poisoning in the United States and other countries. The worst cases of poisoning were in Minamata and Niigata, Japan, where industrial discharge of mercury caused high concentrations in fish and sickness in many people who ate the fish regularly.[40] Methylmercury (see below) concentration in fish from both locations was from 9 to 24 milligrams per kilogram, although some fish had levels as high as 40 milligrams per kilogram.

In Alamogordo, New Mexico in 1969, Amos Huckleby, age fourteen, was the first documented case of mercury poisoning in America. Amos, two of his sisters, and a baby brother fell ill after consuming, not fish, but pork from a pig fed grain treated with a mercury-containing fungicide prior to butchering. Hair samples of the family ranged from 186 milligrams per kilogram of mercury for the father to 2,436 milligrams per kilogram for one of the sisters (the highest level ever recorded for human hair). In the late 1960s, a Long Island woman suffered from symptoms that later were attributed to mercury from eating swordfish daily for nine months on a weight loss diet. Her hair contained 42 milligrams per kilogram of mercury compared with the national average of 2.0 milligrams per kilogram, although by the time of testing she had not eaten swordfish for more than five months.[41] Tests on human subjects in the Minamata area of Japan in the 1950s found hair containing from 300 to 700 milligrams per kilogram of mercury.

The FDA realized that their 0.5-milligram per kilogram action level was overly cautious when ocean fish from areas near no likely pollution often tested higher than 0.5 milligram per kilogram. According to D'Itri and D'Itri (1977), "The FDA's nationwide sampling indicated that 23 percent or 207 million of the approximately 900 million cans of tuna packed in 1970 were above the 0.5-milligram per kilogram mercury action level. The samples represented several species of tuna from both the Atlantic and Pacific Oceans." Swordfish turned out to contain even more mercury than tuna. The swordfish industry was devastated in May 1971 when the FDA advised Americans to stop eating swordfish. Tests on 853 samples of swordfish showed that all but 42 exceeded the 0.5 milligram per kilogram action level. The fallacy of a 0.5 milligram per kilogram guideline was made more obvious when museum specimens of tuna fifty-three to ninty-three years old and a swordfish twenty-five years old were shown to have mercury levels similar to contemporary specimens. In 1979, the FDA maximum safe level was raised to 1.0 milligram per kilogram of mercury based on studies that showed this level adequately protects consumers. The FDA changed the action level again in 1984 to 1.0 milligram per kilogram of methylmercury, because methylmercury is the most toxic form and accounts for most of the mercury in fish. This change had the effect of a further slight raising of the acceptable limit for mercury.

Mercury was found to be widespread in water, stream bottoms, and fish in Georgia in the early 1970s. Because mercury is not very soluble, levels were never very high in water. Yearly average levels in Georgia's rivers ranged from less than detectable to about 1 microgram per liter (part per billion) from the early 1970s to the 1990s, with no apparent trends. However, mercury accumulates in animals that consume contaminated food and in the early 1970s it was widespread in fish and in fish-eating animals. Levels were high enough in fish in some stretches of Georgia streams to cause the Georgia Water Quality Control Board (1971) to warn against consumption of fish caught from those waters. In 1970, two mercury-cell-type chlor-alkali plants were discharging mercury into Georgia streams. Olin Corporation discharged into the Savannah River south of Augusta, and Allied Chemical Corporation discharged into Purvis Creek, which enters Turtle River near Brunswick. Mercury concentrations in fish and other seafood were higher in the Savannah estuary and in the Brunswick area than other areas of the Georgia Coast. Mercury was also higher in the Savannah River below Augusta than in other Georgia streams. The average mercury concentration in largemouth bass from the Savannah River between Augusta and Savannah was 1.5 milligram per kilogram of fresh weight, which was three times the FDA's maximum allowable level at that time. This compared to an average of 0.38 milligram per kilogram in the same species from all of the other fresh water streams tested.

Pollution by mercury has greatly decreased since the first testing began in the early 1970s. The decrease is because of the prohibition of its use in goods like paint, medicine, and batteries, and a great reduction in industrial use in chlor-alkali plants and pulp mills. Mercury was banned from interior paint only in 1990.[42] Mercury use by industry in the United States peaked in

1964 and D'Itri and D'Itri (1977) reported that total use had decreased by one-third by 1975. Agriculture used 570,000 pounds of mercury in pesticides in 1950, but only 106,400 pounds by 1971. Further, USEPA concluded that domestic use of mercury fell by 70 percent from 1980 to 1993. Incidentally, exports of mercury exceed imports and about 78 percent of exported mercury is sold by the federal government, much of it from government stockpiles. Federal sales accounted for 97 percent of the US demand in 1993.[43]

Release of mercury to the environment has probably been decreased to an even greater extent than its use. Allied Chemical Corporation and Olin Corporation, mentioned above, were directed by the Georgia Water Quality Control Board in 1971 to reduce their releases from 3 and 10 pounds per day, respectively, to less than 0.25 pound per day. Operating reports in late 1971 showed that both plants were abiding by the restrictions. Because less mercury is released into Georgia streams there have been decreases in contamination of fish and other seafood since the 1970s. The lower Savannah River and the waters around Brunswick were the two most contaminated areas in the early 1970s. Mercury in seafood species in waters around Brunswick was high in the early 1970s because of industrial discharge of mercury, but it decreased by 42 to 73 percent by the 1990s (Table 4.4). The Georgia Water Quality Control Board (1971) reported that mercury in blue crabs decreased from 1.09 milligrams per kilogram of fresh weight in 1970 to 0.56 milligram per kilogram at the last sampling in 1971. Mercury concentrations in mud in both the Brunswick and middle Savannah River sampling areas also decreased during the same period. It is remarkable that such a drop could occur in one year after reduction of mercury releases, but, in fact, the levels in blue crab near Brunswick in the 1990s are similar to those in late 1971. So, it appears that once mercury pollution was stopped, recovery was rapid.

In most freshwater streams in Georgia, largemouth bass are at the top of the food chain, living almost exclusively on other fish. Therefore, they have among the highest levels of mercury of any species. They are also the most widely sampled fish in surveys for mercury. For these reasons, I have used mercury in largemouth bass to show trends in mercury contamination of freshwater fish. As shown in Table 4.5, mercury in largemouth bass decreased by 81 percent in the Savannah River from Augusta to Savannah since the early 1970s. Mercury in bass in that section of the Savannah River were the highest in the state at 1.55 milligrams per kilogram of fresh weight in 1970–1971, but by 1991–1997 it was only 0.29 milligram per kilogram.

The large decreases in mercury in the Savannah River and the Brunswick estuary were not matched in fish of other Georgia lakes and streams probably because the other streams were not as contaminated, but there has been a decrease in mercury in fish in most waterways since the 1970s. Table 4.5 gives the mercury concentrations in largemouth bass in ten lakes and five rivers sampled in both the 1970–1971 and 1991–1997 surveys. There was an increase in mercury in fish in Lake Seminole and Clarks Hill Lake where mercury in largemouth bass was low in 1970–1971 (0.26 and 0.13

Table 4.4 Mercury in seafood (milligrams per kilogram) in waters around Brunswick, Georgia. The data from 1970-71 are from the Georgia Water Quality Control Board (1971), and the 1991-95 data are from the Georgia Department of Natural Resources (1998a)

Species	1970-71	1991-95	Reduction
	mg per kilogram		%
Blue Crab	1.00	0.58	42
Shrimp	0.22	0.08	64
Flounder	0.66	0.25	62
Croaker	0.45	0.12	73
Spot	0.30	0.10	67

Table 4.5 Mercury in largemouth bass (milligrams per kilogram) in Georgia waters.

Location	Mercury in mg/kg		
	1970-71	1991-97	% change
Savannah river (Augusta to Sav.)	1.55	0.29	−81
Altamaha River (Everett)	0.74	0.24	−68
Ocmulgee river (US Hwy 280)	0.59	0.25	−58
Coosa River (Mayos Bar)	0.42	0.14	−67
Oconee River (Barnett Shoals)	0.41	0.23	−44
Lake Allatoona	0.52	0.20	−61
Lake Harding	0.30	0.13	−60
Lake Lanier	0.52	0.28	−46
Lake Hartwell	0.42	0.24	−43
Lake George	0.40	0.24	−40
Lake Sinclair	0.16	0.13	−19
Lake Jackson	0.20	0.18	−10
Lake Blackshear	0.19	0.16	−16
Lake Seminole	0.26	0.29	+11
Clarks Hill Lake	0.13	0.24	+85
Average	0.45	0.22	−51

The data from 1970-71 are from the Georgia Water Quality Control Board (1971), and the 1991-97 data are from the Georgia Department of Natural Resources, 1998a.

milligram per kilogram, respectively), but in all others there was either no change or a decrease. Even including the two lakes with an increase, the overall decrease in mercury in bass in these waterways, most of which were not highly polluted with mercury, was from 0.45 to 0.22 milligram per kilogram or 51 percent. The 1991–1997 average was similar to that reported by the USEPA (0.27 milligram per kilogram) for largemouth bass in Georgia for the period 1990–1995.[44]

In spite of these data for largemouth bass, the Georgia Department of Natural Resources (1999b) states in its report; "when today's data is compared to data collected from as far back as

1971, it appears that the amount of mercury in most fish in Georgia is about the same." The number of fishing areas with advisories against eating fish increased during in the mid-1990s. The reason was not that mercury levels had increased, but that the level of mercury that triggers an advisory was lowered three-fold.[45] An advisory is now issued to limit consumption of fish when mercury is above 0.25 milligram per kilogram (parts per million or ppm). This contrasts with a statement by the US Food and Drug Administration (2001) that "The 1 ppm limit FDA had set for commercial fish is considerably lower than levels of methyl mercury in fish that have caused illness."

There are differences among Georgia streams in mercury content of fish, and some appear to have natural levels that exceed government guidelines. In surveys of the 1990s, largemouth bass in several smaller rivers arising in the Coastal Plain or southern Piedmont (Canoochee, Ogeechee, Ohoopee, Satilla, Alapaha, Ochlockonee, and Withlacoochee) averaged 1.00 ± 0.22 milligram of mercury per kilogram compared to an average of 0.25 ± 0.15 milligram per kilogram for bass in other Georgia rivers. After a nine-year cleanup of the Ochlockonee River the city manager of Moultrie declared in 1989 that "the Ochlockonee may be one of the few rivers in the area not endangered by mercury levels."[46] However, analyses in 1995 and 1997 showed mercury levels in bass in the Ochlockonee River of 1.04 and 1.30 milligrams per kilogram. In the 1970–1971 survey of mercury in fishes of the state,[47] these rivers listed above were not sampled. Four small samples of three fish species (largemouth bass, bowfin, and shortnose gar) from the Satilla River in Ware County in 1976–1977 had a combined average concentration of 1.72 milligrams per kilogram.[48]

Pollution is not likely the reason for mercury being four times higher in bass in these South Georgia streams, because these rivers are among the least polluted in the state. Higher mercury in fish in these streams probably results from either higher mercury in the soils and sands leached by water entering these streams, or differences in the chemical reactions that bind and release mercury from these soils. Sandy soils in the Coastal Plain and lower pH of streams may result in greater availability of mercury to fish and other aquatic animals. It follows that most of the EPD advisories against eating fish because of mercury "contamination" are for South Georgia streams. Of the sixty-three stream sections that had such advisories in 1998, 70 percent were in South Georgia. Sixty percent were on smaller rivers and streams of that region, which do not have significant pollution problems.

The higher mercury in fish in streams of the Coastal Plain also means that mercury is higher in the wildlife that eat the fish and in predators of the fish-eating animals. River otters, raccoons, and bobcat from the Coastal Plain were found to have higher mercury concentrations in their hair than did those from the Piedmont. The trend for higher mercury in opossum and gray fox in the Coastal Plain was less pronounced, and no differences were observed between the Piedmont and Coastal Plain in mercury in the hair of deer and mink.[49] Osowski et al (1995) later found that mercury was higher in mink in the Coastal Plain than in the Piedmont of Georgia, South Carolina and North Carolina. It is not clear whether mercury in fish-eating animals has decreased since the 1970s along with that in fish. Mercury levels in otter hair and muscle increased from 1976–1977 to 1980–1981, but Clark et al (1981) concluded that this may have been due to a very dry period and higher acidity of waters during the later sampling.

An indirect way of studying the changes in mercury pollution is to examine mercury in sediment deposited in lake beds and estuaries at different times. In a study of sediment in the Savannah estuary, Alexander et al (1994) concluded that there were relatively large inputs of mercury in the 1950s and 1960s followed by a gradual decrease in sediment deposited over the past twenty to thirty years. The burying of mercury in older sediment by more recent deposits makes the mercury less available to aquatic organisms and therefore less harmful to the environment. If, however, these bottom sediments are disturbed by dredging or other development, the buried mercury or other harmful contaminants can be released and pose a danger to aquatic life in areas near the disturbance.

Other heavy metals have shown the same downward trend in concentration in recent years. The US Geological Survey studied contamination of the sediment in lakes in the Apalachicola-Chattahoochee-Flint River basin. They found that zinc, lead, and chromium increased in sediment laid down from the 1930s to the 1970s in Lake Harding near Columbus (Figure 4.11), as well as in other lakes. There was a remarkable decrease in these elements from the late 1970s to 1993. Frick et al (1998) attributed the decrease in heavy metals in the sediment of Lake Harding to the filtering effect

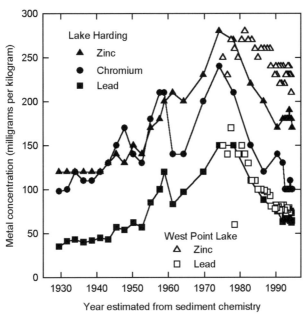

Figure 4.11. Changes in heavy metal concentrations with time in Lake Harding and West Point Lake estimated from analyses of lake sediments. Data from the US Geological Survey website referred to by Frick et al. (1998).

of West Point Lake, which is upstream on the Chattahoochee and was completed in 1974. However, there was a similar decrease in zinc and lead in the West Point Lake sediment during the same period. This means that the decrease probably resulted from less of these two contaminants being released into the river upstream of West Point Lake. Although chromium also decreased in West Point Lake from the late 1970s to 1993 the decrease was less than in Lake Harding. The heavy metals in sediments of Lake Harding and West Point Lake likely came from the metropolitan Atlanta area, since concentrations of lead[50] and other metals were never as high in Lakes Lanier and Blackshear, which are downstream from rural areas.

PCBs are more dangerous industrial contaminants than heavy metals. Before the late 1970s these chemicals were widely used in electrical transformers, lubricants, ink, and sealants. The government restricted their use in 1977 and all uses were cancelled in 1979. PCBs were first monitored in Georgia waters in the 1970s. The danger to humans was not so much in drinking water, but in eating fish and other seafood from contaminated waters. While concentrations of PCB were low in water, they persisted for a long time in lakes and streams and were concentrated in fish in the same way as mercury and some other contaminants. The 1992–1993 report of the Department of Natural Resources states, "In 1976, mixed species of fish sampled from the Coosa River had quantities of PCB's which exceeded the FDA tolerance level of 5 parts per million. Yearly monitoring of fish from the area has shown a dramatic decrease in PCB concentrations over the years. Certain species continue to exceed the revised FDA tolerance level of 2 ppm." Notice that the FDA tolerance level dropped from 5 to 2 milligrams per kilogram (ppm) from the 1970s to the 1990s.

Buell and Couch (1995) studied the presence of PCBs in fish of the Apalachicola-Chattahoochee-Flint River basin from historical (1965–1990) data and in samples they collected in 1992 and 1993.

They reported that PCBs were found in 70 percent of 141 fish samples taken from 1965 to 1990 (Table 4.6). But they found so few PCB-contaminated samples in the 1992–1993 sampling that they stated, "The relative absence of PCB's in the NAWQA (National Water-Quality Assessment) bed-sediment and tissue samples seems anomalous, given their ubiquitous occurrence in the historical monitoring data." However, the lack of widespread contamination of fish with PCBs in this basin was confirmed by data of the Fish Tissue Assessment Projects of the Georgia Department of Natural Resources. The results of the Fish Tissue Assessment Projects have been summarized in Table 4.6. From 1991 to 1997, PCBs were found in only 31.4 percent of 391 samples (Table 4.6) even though the analyses had improved to the point that 0.03 milligram per kilogram of fish could be detected compared to 0.20 milligram per kilogram earlier. The maximum concentration recorded in fish during the 1965–1990 period was 6,060 milligrams per kilogram, but only 1.65 milligrams per kilogram in the 1991–1997 period. It is

Table 4.6 Organic contaminants (milligrams per kilogram) in fish of the Apalachicola-Chattahoochee-Flint River basin in earlier years compared to the 1990s. (Data for 1965 to 1990 from Buell and Couch, 1995; data for 1991-1997 from Georgia Department of Natural Resources, 1998a.)

	Ave. 1965 to 1990[1]			Ave. 1991-1997[2]		
	Median conc.	Max. conc.	%[3] detected	Median conc.	Max. conc.	%[3] detected
	mg/kg			mg/kg		
DDE[4]	0.04	632	55	0.02	6.20	38.6
DDD[4]	0.01	0.20	37			
Chlordane	0.37	857	43	0.07	2.61	13.5
Total PCBs	0.49	6,060	70	0.10	1.65	31.4

[1] From 58 to 239 analyses, some analyses began as late as 1980. [2] 391 analyses [3] Percent of the samples with detectable amounts. [4] Main remaining chemical forms of DDT; summary of these two forms combined for the 1990s.

obvious that PCBs in fish in this large three-river basin no longer exceed the FDA tolerance level of 2 milligrams per kilogram.

PCBs are so persistent in the environment that long-term changes in concentration can be traced in the sediment of the older lakes and in estuaries. PCBs in the sediment of Lake Harding rose from nearly zero in 1930 to about 350 micrograms per kilogram from 1950 through the 1970s, and then dropped to less than 50 micrograms per liter by the 1990s (Figure 4.12).[51] PCBs at one location in the Savannah estuary peaked at about 100 micrograms per kilogram in sediment deposited about 1968 and have decreased to less than 30 micrograms per kilogram since that time.[52] Other locations followed the same trend, but concentrations were never as high. Even though PCBs can still be detected in Georgia lakes and fish, it is apparent that their presence and concentration has decreased during the last quarter of the twentieth century.

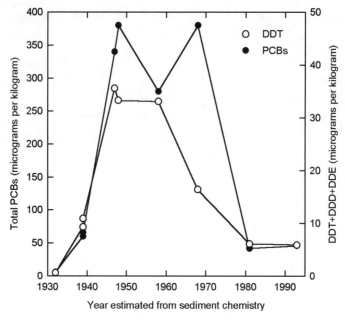

Figure 4.12. Changes in the residues of DDT and PCBs in sediment of Lake Harding representing years from 1932 to 1993. Data from the US Geological Survey website referred to by Frick et al. (1998).

More concern has been expressed about contamination of water and wildlife with pesticides than most other pollutants. The focus of Rachel Carson's book, *Silent Spring*, which caused so much controversy in the 1960s, was the damage pesticides caused to the environment. As we shall see, however, the danger to the environment is much less than it used to be because there are less pesticide residues. One of the reasons that Carson's book caused such a controversy was the persistence of pesticides, especially dichloro-diphenyl-trichloroethane (DDT), and their concentration in certain birds and other wildlife. Soon after the publication of *Silent Spring*, there was a concerted effort to monitor pesticide residues in waters and wildlife. There was a push to ban DDT and other organochlorine compounds used as pesticides. DDT was banned in 1973. Chlordane, heptachlor, dieldrin, and related compounds were used agriculturally through 1974, but through the 1980s as treatments for termites.

Those pesticides that were banned in the 1970s and 1980s have disappeared or are slowly disappearing from the environment. The persistence for which they were banned also keeps them preserved in some places. For example, DDT and the chemicals formed when it breaks down can still be found in sediment buried in lakes and slow moving streams. Sediment in Lake Harding contains over 30 milligrams per kilogram in the layers that were deposited in the 1950s, but deposits of the 1980s and 1990s contain only about 5 milligrams per kilogram (Figure 4.12). In the Savannah River estuary, where as much as 15 to 20 milligrams per kilogram of DDT residues occur in sediments deposited in decades past, less than 5 milligrams per kilogram are in sediment deposited in the 1990s.[53] The burial of DDT and other dangerous and persistent pesticides in sediment of lakes and streams makes them less available to fish and other aquatic organisms. This is reflected in the lower residues in fish in recent years than in the past. The DDT breakdown products found in highest concentration in fish are DDD and DDE. DDE was found in 55 percent of the fish samples from the Apalachicola-Chattahoochee-Flint River basin during the period, 1965–1990,[54] but both forms were found in only 38.6 percent of the samples from 1991–1997.[55] The difference in maximum concentrations was even more significant, being 632 milligrams per kilogram for DDE in the period 1965–1990 and only 6.20 milligrams per kilogram for the two chemicals in 1991–1997, a 100-fold decrease. The sample with 6.20 milligrams per kilogram and the only samples with DDE or DDD above 0.45 milligrams per kilogram were from the Albany By-pass Pond. DDT itself was found in only 12 percent of the fish samples from 1965 to 1990, but in none of the samples in 1991–1997. The maximum concentration allowed by the FDA in meat is 5 milligrams per kilogram. So it is clear that DDT in fish from Georgia waters is no longer a danger to humans.

DDT and other pesticides were banned, in part, because of their danger to wildlife. "Peregrine falcons and bald eagles were judged to be at the brink of extinction in the 1960s, the victims of DDT and other pesticides that caused reproductive problems and weakened eggs."[56] By 1998, the peregrine falcon had made such a recovery in the US that Seabrook wrote, "From the summit of Stone Mountain, Interior Secretary Bruce Babbitt is expected to announce today that the federal government is taking the rare step of removing a bird—the peregrine falcon—from the list of endangered species." Seabrook went on to say, "Several pairs of falcons are now seen regularly in Georgia, including a pair that nests each spring on ledges in skyscrapers in downtown Atlanta." The bald eagle has also been removed from the endangered list.

Other organochlorine pesticides have also decreased in wildlife and in lake sediments. Chlordane detection decreased from 43 percent of the fish samples taken from the Apalachicola-Chattahoochee-Flint River basin in the 1965–1990 period to 13.5 percent of those taken from 1991–1997 (Table 4.6). This was in spite of the fact that the detection limit had decreased from 0.08 to 0.03 milligram per kilogram. The maximum concentration detected in the earlier period was 857 milligrams per kilogram compared to 2.61 milligrams per kilogram from the 1991–1997 period. Dieldrin in the three-river system was detected in only 1.6 percent of the fish samples taken from 1991 to 1997 and the other organochlorine pesticides were detected in less than 1 percent of the samples or none at all. Chlordane decreased in sediment of Lake Harding from values of 46–84 micrograms per kilogram in the 1950–1970 period to values below 10 micrograms per kilogram in two samples taken after the 1970s. Because of later use as a termite treatment in the Atlanta metropolitan area, concentrations in sediment of West Point Lake decreased to a lesser extent, remaining in the 20–40 micrograms per kilogram range into the 1990s.

Persistence was one of the desirable properties of the organochlorine pesticides, because it meant that control of the pest would last for a long time after application of the chemical. The controversy surrounding the dangers of these pesticides spurred efforts to develop pesticides that were less persistent. That effort resulted in many pesticides that dissipated in a short time. Today, although pesticides are regularly used in agriculture, industry, public facilities, and even homes, residues are seldom a problem. Surveys, conducted more frequently today than in the past, seldom find pesticide residues in concentrations that are dangerous. Frick and Crandall (1995) found trace concentrations of six pesticides in water samples from 44 percent of wells adjacent to farm fields in the Apalachicola-Chattahoochee-Flint River basin, "but none exceeded drinking water standards." Frick (1997) found during a 1994–1995 survey of twenty-one wells and nineteen springs in the Chattahoochee River basin, that dieldrin was the most commonly detected pesticide although it was canceled for agricultural use in 1974 and for treating termites in 1987. Although dieldrin was the most widely distributed of the pesticide residues, it was not found in concentrations above the USEPA guidelines. Diazinon, an insecticide widely used today, was found in only one sample out of 199 in a concentration (2.8 micrograms per liter) exceeding the USEPA lifetime drinking water advisory (0.6 microgram per liter).

It will be surprising to many that the most likely source of pesticide pollution is urban rather than agricultural. Six Georgia and Florida streams sampled intensively in 1993 and 1994[57] had a total of twenty-five of the forty-seven pesticides tested. Only one of the 217 samples exceeded drinking water standards for only one pesticide, the widely used herbicide, simazine. That was in Sope Creek in metropolitan Atlanta. However, that stream and another urban creek in Tallahassee, Florida, had concentrations of from one to three insecticides that exceeded guidelines for protection of aquatic life during thirteen of the fourteen months of sampling. Four agricultural streams exceeded the same guidelines for only three months or less. In the Chattahoochee and Flint River basins, chlordane was primarily detected in the rivers and lakes downstream from Atlanta, Columbus, and Albany.[58] A similar conclusion was reached by Stell et al (1995) for the distribution in the same river system of chlordane, heptachlor, and dieldrin, all organochlorine insecticides.

The data in Table 4.7 is a summary of the most frequently occurring pesticides found in an extensive survey of creeks in the Apalachicola-Chattahoochee-Flint River basin.[59] Twelve agricultural stream sites in southwestern Georgia along with seven each of suburban and urban sites were sampled for forty-seven pesticides from 1993 to 1995. The urban and suburban sites sampled were on creeks in the metropolitan Atlanta area, except one on Bull Creek at Columbus. A total of 28 pesticides were found in the 133 samples from agricultural streams. However, twenty-nine pesticides were detected in urban creeks from only thirty-six samples, and thirty-six from ninety-two samples from suburban creeks. Insecticides were detected almost exclusively (thirty of thirty-two detections) in stream water samples from suburban and urban sites. Average concentrations of seven of the thirteen pesticides listed in Table 4.7 were higher in either the urban or suburban creeks or both, compared to those in agricultural areas. Two of the pesticides with higher concentrations in agricultural streams (prometon and chlorpyrifos) were found in only 1 percent of the samples, whereas they were detected in from 12 to 78 percent of suburban and urban samples. Although levels of the pesticides tested seldom exceeded the

Table 4.7 Concentration (micrograms per liter) and frequency of pesticides in streams of the Apalachicola-Chattahoochee-Flint River basin — 1992–95. From Frick et al., 1998.

	Agriculture		Suburban		Urban	
	Ave. conc.	% of[1] samples	Ave. conc.	% of samples	Ave. conc.	% of samples
Alachlor	8	18	23	3	—[2]	0
Metolochlor	16	64	14	9	8	22
Atrazine	29	58	49	96	49	92
Metribuzin	21	3	10	1	10	14
Simazine	82	5	249	97	127	100
Trifluralin	35	14	8	12	14	25
Tebuthiuron	26	2	33	60	281	89
Prometon	220	1	181	12	43	36
Diuron	—	0	158	6	363	50
Pendamethalin	8	2	62	37	101	36
Malathion	8	1	32	11	46	25
Diazinon	12	<1	43	90	126	94
Chlorpyrifos	99	1	14	47	31	78
Sites sampled	12		7		7	
No. pesticides found	28		36		29	
No. samples tested	133		92		36	

[1] Percentage of samples in which the pesticide was found. [2] Blank spaces indicate the pesticide was not found in those locations.

USEPA standards for drinking water, dieldrin exceeded the standards in one or more samples from five of thirty-seven sites in metropolitan Atlanta in 1994 and 1995. The greater prevalence and concentration of pesticides in urban and suburban streams may result from more frequent and heavy use for lawns, gardens, households, and public areas. On the other hand, the higher percentage of impervious land surfaces in urban areas may result in more runoff of pesticides applied.

Even though water in urban areas is more likely than rural streams to have pesticides, the occurrence and danger of pesticides have greatly decreased all over Georgia. Impairment of Georgia's waters results from pollutants other than pesticides. There were 2,385 miles of streams and 64,813 acres of lakes and estuaries listed by the Georgia Department of Natural Resources (2001a) in 1998–1999 as not supporting "designated uses,"(swimming, fishing, etc.). Pesticide contamination was not listed as the reason any of these waters failed to support their "designated uses."

Overall Improvements in Water Quality

Judging the change in quality of stream and lake water is complicated by the several measures of water quality, one of which may improve, while others remain the same or even become worse. Several of the more important measures of water quality are biochemical oxygen demand, dissolved oxygen, fecal coliform bacteria, nitrogen, phosphorus, specific conductance (a measure of dissolved substances), and organic carbon. In a recent study Peters and Kandell (1997) evaluated eighteen sites around Atlanta using a water quality index (WQI) made up of a combination of these individual measures. The sites ranged from Allatoona Dam on the Etowah River north of Atlanta to Fairburn on the Chattahoochee River south of the city, and the Flint River near Lovejoy. The individual measures of quality were expressed as fractional percentile ranks, ranging from zero to one, with zero denoting the best (least pollution) and one denoting the poorest water quality. These rankings were then combined to form the water quality index for two periods (1986–1989 and 1990–1995).

The water quality index improved at each of the eighteen sites from 1986–1989 to 1990–1995, with improvements ranging from 6.7 to 29.0 percent. The improvement near the dam in Lake Allatoona was 15 percent. This is contradictory to an article in the *Atlanta Journal-Constitution*[60] which cited a study by Kennesaw State University that Lake Allatoona is "in danger of being dead in 10 years." The average increase in water quality for the eighteen places tested was 18.5 percent between the periods 1986–1989 and 1990–1995. Such a decrease in water quality index over this relatively short time represents a significant overall improvement of water quality. It shows that there is continued improvement in the region of the state where water quality is poorest, even though the greatest improvements were in the 1970s.

Eutrophication of lakes is also judged by different measurements, which may or may not be closely tied to each other. The excessive growth of algae is the result of too much nutrients. The algae die, settle to the bottom, and decompose, using up oxygen, which results in low concentrations near the bottom of the lake. So, the eutrophication of a lake may be judged by nutrient levels, concentrations of algae (usually measured as chlorophyll), clarity of the water, oxygen concentration, or other characteristics. Phosphorus and nitrogen from sewage disposal and from nonpoint sources have been considered major causes of eutrophication and the algal blooms in lakes that result from too much of these nutrients. Apparently, some addition of nutrients is desirable for production of fish, because fertilization of farm ponds is recommended.

The Department of Natural Resources, Environmental Protection Division, calculated a Trophic State Index for twenty-seven lakes in Georgia for the years 1984 to 1993, using a combination of chlorophyll *a*, water clarity and phosphorus.[61]

Using that index it is clear that the condition of most Georgia lakes has not changed much since 1983 (Figure 4.13). While Blue Ridge Lake became more eutrophic since the mid 1980s, Lake Jackson has become much less eutrophic since the late 1970s. Eutrophication in Lake Harding also appeared to decrease somewhat in the early 1990s. West Point Lake does not appear to have changed much since soon after its construction in 1974, even though there was great concern that it would rapidly become eutrophic because of the nutrients from Atlanta's waste. The two separate sets of data for West Point Lake do not coincide very well, probably because I calculated the earlier one from chlorophyll concentrations published by Bayne et al (1990) and the second was calculated from chlorophyll, phosphorus, and transparency measurements as indicated above. Although the later one indicates less eutrophication, there

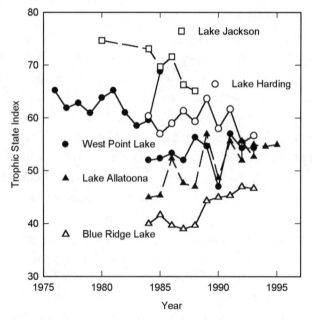

Figure 4.13. Changes in the eutrophication status of five Georgia lakes during three decades. Lake Jackson (Kamps, 1989), West Point Lake 1976–1985 (Bayne et al., 1990); West Point Lake, Blue Ridge Lake, Lake Harding, Lake Allatoona 1984–1993 (Georgia Department of Natural Resources, 1997); Lake Allatoona 1992–1995 (Burruss Institute of Public Service (1999)

does not appear to be an up or down trend since 1976. A review of water quality in Lake Lanier in 1966, 1973, and 1991 revealed that although phosphorus had increased, for total suspended solids, turbidity, and secchi depth (clarity) "No clear long-term changes are apparent...."[62]

The dangers of water pollution have greatly diminished in urban and rural areas since the times when sewage was untreated, erosion was unchecked, and persistent organochlorine insecticides were widely used. As the data discussed in this chapter show, however, most of the problems that remain are related to the concentration of people and industry in metropolitan centers. One of the telling signs that the situation has changed for the better, is the emphasis in many recent scientific articles and in the popular press on the number of detections of contaminants rather than of levels exceeding government standards (note the discussion of Tables 4.5 and 4.6). Reduction in detection of contaminants with time is the more remarkable because modern analytical methods has made it possible to detect ever smaller concentrations. All of the average concentrations listed in Table 4.7 are less than 1 milligram per liter (ppm).

Our drinking water is probably safer now than at any time in the twentieth century. This seems contradictory to the trend toward increased consumption of bottled water. Preference for bottled water must result from the perception of greater safety, better taste, or convenience. Whatever the reason, there is little doubt that perceptions about tap water quality lags behind reality. News about water quality in newspapers and on television is usually about accidents or government inaction that results in lowering of water quality. The improvement of water quality by government regulation, changing land use patterns, or industrial remediation is treated as old news, not worth reporting. Many of the articles about water quality advocate a course of action or highlight problems of existing systems, rather than objective analysis of problems and progress. It is little wonder that the public mistrusts public water supplies.

Even surveys by scientists about water are not always objective. Jordan and Elnagheeb (1993) mailed a survey to 567 Georgia residents in 1991 asking about their willingness to pay increased water bills or service charges for improved water quality. They prefaced the survey with an introduction, which included the following statements. "The Environmental Protection Agency (EPA) has ranked the state of Georgia as second in the nation for *potential* (my emphasis) contamination of underground water." "On the other hand, if agricultural practices did not change, the amount of nitrates in the groundwater would increase. So the costs of cleaning water from nitrates will go up." These statements not only biased answers to the survey, but ignored the fact that a tiny fraction of water in Georgia approaches the drinking water limit of 10 milligrams per liter of nitrates, and there is no evidence that the fraction is increasing. The response to the survey was 192 questionnaires returned; 150 from households on city/county water systems and 42 using water from private wells. Of the 150 respondents who were on city or county water systems, 27 percent rated their drinking water as "poor" and 23 percent were "uncertain" about their water quality. Results were not given for the other rating categories, "very safe," "safe," and "fair." It is likely that the preface to the questionnaire raised concerns about water quality and defeated the purpose of the survey.

The limits of polluting substances in streams and lakes have been set at conservative levels to protect human health. The levels are at least somewhat arbitrary, because experiments cannot be performed on humans to establish "safe" levels of contaminants. In many cases, industry and other interest groups have complained about "safe" limits being set too low. However, such groups are not alone in complaining about the strict limits on stream pollution.

The Georgia Department of Natural Resources (1976) complained that:

> It is further estimated that some 5% of Georgia's streams cannot meet the water quality criteria for swimming or fishing due to natural conditions. These waters include primarily the swamp-like waters of South Georgia, which exhibit naturally low dissolved oxygen, low pH (high acidity), and high water temperatures during summer and fall months. The fact that these natural waters in South Georgia and other parts of the State do not meet fishing and swimming criteria certainly does not mean that they are not fishable and swimmable. People have recreated in certain of these waters for years, and fish have thrived in these streams for thousands of years.

Later in 1993, referring to new, more stringent fecal coliform regulations, the Georgia Department of Natural Resources stated about streams: "EPD (Environmental Protection Division) and local government monitoring programs have documented levels of fecal coliform bacteria in excess of the new standards...even in areas not extensively impacted by man such as in national forests." Again in 1997, the Georgia Department of Natural Resources explained that 30,000 of the 92,000 acres of estuaries on the coast restricted for oyster harvest in 1995 "have the classification because of bacteria levels which occur naturally as a result of local wildlife."

So the perplexing possibility is raised by this state agency that some officially designated pollution results from natural causes. In fact, in local situations, "pollution" by wildlife is substantial, as in the Lake Acworth incident mentioned earlier. An instance of pollution of Macks Island in the Okefenokee Swamp by nesting wading birds, mostly white ibis, was described by Stinner (1983). It was estimated that the 5,000 to 30,000 birds that nest there deposited the equivalent of 800 to 5,000 pounds of dry manure per acre per year in the rookery, and raised the phosphorus concentration in the water to 0.57 milligram per liter (five times the suggested limit for streams). Stinner concluded "Therefore, breeding colonial wading birds create a local natural eutrophication in an otherwise nutrient impoverished swamp habitat." While it is playing with words to suggest that such natural fertilization is pollution, the effects on plants and animals that inhabit the water are similar whether the source is natural or manmade.

Protection of Stream Quality

A number of characteristics offset the polluting potential of waste put into streams or on the land near streams. For many pollutants the streams are self-cleaning. Solids settle to the bottom some distance below the place they enter a stream, especially if water movement is slow. Bacterial pollutants die with time if they have no substrate for growth. As shown in Figure 4.2, fecal coliform in the Chattahoochee returned to very low levels in a short distance below metropolitan centers. Phosphorus followed a similar pattern. It may be that suspended sediment helps remove phosphorus and perhaps other contaminants from streams and lakes. Water flowing from Flat Creek into Lake Lanier was clearer and lower in total phosphorus and sediment the further into the lake and away from the mouth of the creek water samples were taken.[63] It was concluded that total phosphorus was low in Lake Lanier and perhaps in other southeastern lakes because clay particles adsorb it strongly and settle out of the water. The self-cleaning function of streams is one of the main reasons that stream pollution is not more widespread. It is also the reason that most streams return to healthy conditions soon after the source of pollutants is removed. Of course, the more resistant the pollutant is to breakdown or removal the more slowly the stream returns to normal. Mercury and organochlorine compounds like DDT are exceptional pollutants because they are so resistant to removal from the environment.

The burial of persistent pesticides, PCB's, and heavy metals under sediment in stream and lake bottoms removes them from contact with most living things, and is, to some degree, safe storage. At least they cause less damage than if they were still in the water. But these buried contaminants are a long-term problem akin to the stream bank deposits of sediment from soil erosion. Those persistent pollutants in stream bottoms are subject to migration down the watersheds with the sediment that covers them; those in lake bottoms are likely to stay buried unless disturbed by operations such as dredging. The pesticides and other organic contaminants will eventually decompose and become harmless, but the heavy metals are likely to stay in the sediments until they are disturbed. The good news is that we have quit putting these dangerous chemicals in the streams.

There is much concern about the loss of wetlands in Georgia, but among the lamentations about loss of wetlands, one seldom hears the gain in benefits of ponds and lakes. These reservoirs act as filters for streams. When stream water enters a reservoir, the velocity slows and materials suspended in the moving stream settle to the bottom. In addition to the settling of solids, reservoirs that are large in relation to the flow of the streams that feed them hold the materials for a long time, allowing more time for the settling and dissipation of pollutants before the water flows past the dam. One of the benefits cited for the proposed construction of West Point Lake on the Chattahoochee River[64] was that it would sufficiently clean the river of bacterial contamination to allow recreational use of the river below the lake. Indeed, fecal coliform bacteria in the Chattahoochee River at West Point (just below the lake) in the 1990s are only 1 or 2 percent of the numbers in the late 1960s and early 1970s, before the lake was formed. The improvement is partially due to sedimentation in the lake, but in large part to better sewage treatment in cities upstream.

Improved sewage treatment in the Atlanta area did not clean the Chattahoochee of nitrogen and phosphorus to the same extent as it removed bacteria. However, much of the nutrient load in the Chattahoochee River south of Atlanta is utilized by algae or settles out in West Point Lake and to a lesser extent in Lake Walter F. George and Lake Harding.[65] Lakes and ponds provide most of the same benefits attributed to wetlands. In addition to their filtering function, they attract wildlife, especially fish and the birds and animals that prey on them; they are likely to contribute to the recharge of underground water reservoirs; and they are important in flood control. Lakes and ponds also provide recreation to a much greater extent than wetlands. Construction of reservoirs has undoubtedly contributed to quality of water in the state and has certainly offset much of whatever harm was caused by loss of natural wetlands. As discussed in chapter 3, there has been a great increase in the numbers of ponds and lakes since the 1950s.

The filtering action of lakes and ponds has long been known and the regulations that now govern construction projects in Georgia recognize the benefits. Certain construction sites are now required to have settling ponds downstream of the construction to filter out pollutants, mainly sediment. But sewage systems and on-farm treatment of animal wastes have used "settling" or "holding" ponds for years to clean water before it is released into streams. Even simple waterways like canals remove much of the nutrients that pass through them. Van Kessel (1977) found that sewage-contaminated water flowing in an 800-meter-long (about one-half mile) canal lost 56 percent of the nitrate that entered the canal. It was concluded that the disappearance of nitrate was caused mainly by denitrification (conversion to the gaseous form—N_2) in the sediment.

There are methods that can be used to remove minerals not removed in sewage treatment. Municipal wastewater was irrigated on land planted in pines from the Clayton County sewage treatment system beginning in 1982.[66] Sections of the area were irrigated once a week with 2.5 inches of wastewater. The heavy irrigation was used to saturate the soil and promote loss of nitrogen by denitrification. Nitrate and phosphorus were measured before 1982 and afterward in shallow wells downslope from the irrigated area to determine if nutrients were moving into the ground water. Concentrations of nitrate and phosphorus, and the heavy metals, lead, cadmium, and mercury were actually lower in downslope wells on the irrigated areas than in wells placed outside of the area irrigated. Concentrations were no different before and after irrigation began. The conclusion was that the pine-forested area effectively removed these nutrients and heavy metals from the wastewater before it entered streams.

Vegetation bordering streams protects them from the entrance of many pollutants that are put onto the land. This is likely one reason that pollution from agricultural fertilizer and pesticides is so rarely found. When there is a forested zone, especially wet forest, between fields and water, sediment, fertilizer, and pesticides washed from the fields are trapped in the forested zone. After study of a Coastal Plain watershed, Hubbard et al (1990) stated that:

> Overall, the study showed that, as measured on these watersheds, Coastal Plain streamflow is of good quality in terms of both dissolved and suspended solids. This good quality may reflect land use practices designed to prevent soil erosion, but primarily reflects the Coastal Plain landform shape, which causes sediments eroded from the uplands to be deposited in the riparian zone before they can enter streamflow.

In another study, 82 percent of water-borne N, 54 percent of P, and 42 percent of calcium were filtered out by the stream-bank ecosystem.[67] The removal of these nutrients before water entered the streams was accounted for by denitrification and uptake by vegetation. Similar claims for removal of pollutants are made for wetlands. In fact, many of the forested stream-bank areas are classified as wetlands. Following on from its success in using forested areas to remove nutrients and heavy metals from municipal wastewater, Clayton County has developed plans to use 660 acres of wetlands for the same purpose.[68]

In summary, Georgia's water is much cleaner at the end of the twentieth century than it was at the middle. One of its most polluted streams, the Chattahoochee River, was improved enough even by 1980 that the USEPA declared "The poet Sidney Lanier, who glorified the river in his 'Song of the Chattahoochee' in 1877, would once again be proud of his river."[69] The improvement is the result of many factors, from government regulations to changes in land use. The concern of the public for clean water was a driving force for government action. Better farming practices and development of safer agricultural chemicals have played a large role in cleaning the water. Increasing concentration of people in metropolitan areas created many of the water quality problems that existed in mid-century and the concentration of people has increased. However, improved technology, better economic conditions, and public awareness have more than overcome the problem. The public needs to be aware of the progress that has been made, so that it has confidence in the safety of the water and in the ability of government to solve problems of the environment.

[1] US Environmental Protection Agency, 1980.
[2] Wetherington, 1994.
[3] Georgia Statistical Abstract, various years.
[4] Tarr, 1994.
[5] Callahan et al, 1965.
[6] Georgia Department of Natural Resources, 1990a.
[7] Kuhn et al, 1990.
[8] Tarr, 1996.
[9] Wetherington, 1994.
[10] US Bureau of the Census, 1920–2000.
[11] Campo, 1999.
[12] See chapter 3.
[13] Oppenheim, 1996.
[14] Meade, 1976.
[15] Georgia Water Use and Conservation Committee, 1955.
[16] Campo, 1999; Soto, 1999.
[17] US Department of the Interior, 1966.
[18] US Department of the Interior, 1970.
[19] Troxler et al, 1983.
[20] Plummer, 1997.
[21] US Department of the Interior, 1966.
[22] US Department of the Interior, 1969.
[23] Bayne et al, 1983.
[24] Avery, 1999.
[25] Odum and Turner, 1987.
[26] Frick et al, 1996.
[27] Patrick et al, 1992.
[28] Tyson et al, 1995.
[29] Shellenberger et al, 1996.
[30] Gould, 1995.
[31] Frick and Crandall, 1995.
[32] US Bureau of the Census, 1920–2000.
[33] Patrick et al, 1992.
[34] Patrick et al, 1992.
[35] Peters et al, 1997.
[36] US Geological Survey, 1999.
[37] Burruss Institute of Public Service, 1999.
[38] D'Itri and D'Itri, 1977.
[39] Leigh, 1997.
[40] D'Itri and D'Itri, 1977.
[41] D'Itri and D'Itri, 1977.
[42] Georgia Department of Natural Resources, 1998a.
[43] US Environmental Protection Agency, 1997.
[44] US Environmental Protection Agency, 1999b.
[45] Georgia Department of Natural Resources, 1999b.
[46] Associated Press, 1989.
[47] Georgia Water Quality Control Board, 1971.
[48] Halbrook et al, 1994.
[49] Cumbie, 1975; Clark et al, 1981.
[50] Callender and van Metre, 1997.
[51] Frick et al, 1998.
[52] Alexander et al, 1994.

[53] Alexander et al, 1994.
[54] Buell and Couch, 1995.
[55] Georgia Department of Natural Resources, 1998a.
[56] Seabrook, 1998a.
[57] Hippe et al, 1995.
[58] Buell and Couch, 1995.
[59] Frick et al, 1998.
[60] Reinolds, 2000.
[61] Georgia Department of Natural Resources, 1997.
[62] Xiao-Qing and Rasmussen, 1999.
[63] Mayhew and Mayhew, 1993.
[64] US Department of the Interior, 1966.
[65] Frick et al, 1996.
[66] Gaskin and Nutter, 1989.
[67] Todd et al, 1983.
[68] Atlanta Constitution, 2000.
[69] US Environmental Protection Agency, 1980.

CHAPTER FIVE

Clearing the Air

"And we used lots of ultra-violet [sic] light in our office on little children because they got no sunshine. It was so smogged-in with smoke because everybody burned soft coal. All the factories burned soft coal."

—Dr. Leila Denmark
about Atlanta air in the 1920s and 1930s[1]

Smoke and Smog

Air pollution is not a new circumstance. It apparently affected even prehistoric people. According to Brimblecombe (1987), "Blackening of lung tissues through long exposure to smoky interiors appears to be the rule rather than the exception in ancient remains, regardless of whether they originate from polar or from tropical regions." Outdoor air pollution was a problem even in old England. King Edward I banned the use of "sea coal" (coal collected from beaches of England) in the early fourteenth century because of the noxious fumes it caused in the city of London. Coal-smoke pollution became so bad that John Evelyn wrote in 1661, "It is this horrid smoake, which obscures our churches and makes our palaces look old, which fouls our clothes and corrupts the waters so that the very rain and refreshing dews which fall in the several seasons precipitate this impure vapour, which with its black and tenacious quality, spots and contaminates whatever is exposed to it."[2] In a book entitled The Big Smoke (a nickname for London), Brimblecombe (1987) describes problems of smoke and fog in London going back to the Middle Ages. Efforts were made even back then to reduce the use of coal and keep brick kilns and other polluting industries away from the city. Death rates were correlated with the great "fogs" of the late seventeenth century, a period when coal imported into London reached about 300,000 tons. When the first measurements of sulfur dioxide were published in England in 1895, they were 50 to 100 times higher than the highest yearly averages measured in Georgia (in the 1970s).

For many in the United States, the vision of pollution is smoke pouring from factory and power plant smokestacks, symbols of the industrial age. We remember the pictures of factories in cities like Chicago, Pittsburgh, and Newark, pouring forth smoke and not only blackening buildings and city streets, but also casting a pall of smoke and smog across the city skyline. Almost forgotten now, however, are the famous smog episodes in the Meuse Valley of Belgium (1930), London (1952), and Donora, Pennsylvania (1948). In the Meuse Valley episode, sixty people died in the first week.[3] Sulfur dioxide concentrations there from the burning of coal were in the range of 9,500 to 40,000 microliters per cubic meter (compare later to the highest yearly average in Atlanta of about 16 microliters per cubic meter). In Donora, twenty people died and 1,440 were seriously ill.[4] Donora was an industrial town of 13,000 population in 1948, 30 miles south of Pittsburg. One description of the conditions in Donora on Wednesday, 27 October, the day the smog started was vivid, "It was reported that streamers of carbon appeared to hang motionless in the air and that visibility was so poor that even natives of the area became lost." The people and the air of the area were studied for several years after that smog episode had passed. In 1966, eighty people died in New York City during a four-day smog and Governor Rockefeller declared a state of emergency. Such occurrences have disappeared and urban scenes have become much lighter and less drab with time because the air has become cleaner.

When we read about the federal government withholding highway funds because ozone is too high in Atlanta air, we may be convinced that air quality is getting worse. Commuters who see the haze in Atlanta on a hot, still day may believe that air pollution is at its worst ever. But it is not. Believe it or not, Atlantans can breath easier now than forty to fifty, or even twenty to twenty-five years ago. Asthma and other respiratory diseases are unlikely to be worse than earlier because of pollution, in spite of recent newspaper articles to the contrary. Air-pollution-related illnesses in mid-century were likely attributed to something else or accepted as without external cause. We do not know how foul the air was in Georgia more than fifty years ago because no measurements were made. Air quality in Atlanta was almost certainly not as bad as it was in some of the larger, more industrial cities, but it was worse than now. In the last half of the twentieth century pollutants have been reduced remarkably, not only in Atlanta but also across the United States. Those who characterize the air in Atlanta as bad and getting worse may conclude that the kinds of pollutants in the air now are more dangerous than those at mid-twentieth century, but it is unlikely speculation. In any case it is difficult to argue against data that show most pollutants have been reduced, not marginally, but 50 percent or more.

The improvement of Georgia's air quality is the more remarkable because our population has more than doubled from 3.4 million people in 1950 to 8.2 million at the century's end and our standard of living has increased even more. The number of motorized vehicles and power driven devices has increased beyond our realization. Nearly all of this power is derived from the burning of fossil fuels, which produces air pollutants. The clearing of Atlanta's air was achieved while the number of cars almost quadrupled in the thirteen-county metropolitan area, from 705,000 in 1965 to 2,695,656 in 1995.[5]

In the decade after the Donora "smog," concern developed about unhealthy air all across the United States. The US Public Health Service began air quality measurements in many locations around the country, including four in Georgia. However, in those days there was little understanding about the substances in the air that caused health and environmental problems. The most obvious problems were the dust, dirt, and smoke that urban populations had to endure; the dust that settled on store windows, the soot that speckled white laundry still on the line, and the pall of gray-black that hung in the morning sky in factory districts on still days. So, the measurements made in the 1950s by the US Public Health Service were of particles suspended in the air.

The focus shifted with time from smoke and dust to invisible contaminants such as sulfur dioxide, the principle component of acid rain; carbon dioxide, the main cause of predicted global warming; and ozone, the modern bane of respiratory patients. In 1976, for example, there were at least thirty stations in Georgia measuring particles in the air, but only one measuring ozone (Georgia Perimeter College, Decatur; formerly Dekalb Junior College); in the late 1990s, there were eighteen to twenty stations for each of these pollutants. There was even a time in the late 1950s and 1960s when the greatest worry about air pollution was fallout from nuclear testing. Of course the worry about nuclear fallout was contamination of not just air, but of water, food, and whatever else radioactive elements settled on. Concern about nuclear contamination now seems faded and far away.

How Clean Was the Air Back Then?

The standard for clean air for many people is the clear and sunny skies associated with the countryside. Although we do not have much scientific evidence of the clarity of either countryside or downtown air in the early- to mid-twentieth century, it most likely was dirtier than now, especially in the towns and cities. Even as far back as the 1860s there was suspicion about the unhealthy cities. One Georgian, extolling the virtues of clearing and cultivating land, is quoted in the Southern Cultivator magazine as saying, "Shall we build cities while the forest is untouched by man?... Shall we erect factories, those parents of hollow cheeks, and sunken eyes and hectic coughs and short lives" while the cheerful land "invites us to healthful and remunerative labor?"[6] There is information about the situation early in the twentieth century that causes suspicion about the clean air that we envision for the old days. The use of soft coal for heating homes, running trains, and powering factories put so much black smoke in the air that it was not only a nuisance, but a health hazard. In the early 1900s coal was the most used fuel in the United States. Consumption dipped during the depression of the 1930s, but in the 1920s and 1940s it was roughly the same for the United States as in 1975 at 550 million tons per year.[7] As late as 1940, 69 percent of urban households in Georgia heated with coal and 18 percent heated with wood.[8] The coal that Georgia burned came mostly from the regions where the highest-sulfur coal is mined. There was no restriction early in the century on sulfur content of the coal, and no scrubbers on smokestacks to remove pollutants.

It is probably safe to say that more people died from air pollution in Atlanta in the 1930s than in the 1990s, even though the population in the 1930s was only one-third of that in the 1990s. The following revelation from Dr. Leila Denmark,[9] one of Atlanta's first women physicians, says a great deal about air quality in Atlanta in the 1920s and 1930s.

> When I moved to Atlanta [in 1926], by ten o'clock you had a mustache. They all burned soft coal, all the heat they had was soft coal. There was so much smoke that you'd inhale it and your upper lip would be black. And we used lots of ultra-violet [sic] light in our office on little children because they got no sunshine. It was so smogged-in with smoke because everybody burned soft coal. All the factories burned soft coal. The trains puffed in—a terrific amount. We burned leaves when I moved here, we'd pile them up and set them on fire. You couldn't see across the street. We talk about environmental contamination, you could smell sulfur all over the place. It was really bad.... And you'd see an autopsy at Grady Hospital, or any place you'd see the autopsy, if the person was a pretty good age, the lungs were striped. It looked like a zebra. They'd inhaled so much carbon that the lung was black, had black streaks in it.[10]

The ultraviolet light that Dr. Denmark speaks about was used to supply children with vitamin D. Ultraviolet light, from lamps or sunshine, converts sterols in the skin to vitamin D. Now, of course, milk and other foods are fortified with vitamin D.

The situation probably did not improve before an article appeared in the *Atlanta Journal* (1949) describing a study linking air pollution to respiratory illnesses in several American cities, including Atlanta. The article began with a conclusion of the scientists that Atlanta's air was polluted to such a degree that it bears " a direct and significant relationship to death rates for pneumonia, pulmonary tuberculosis and cancer of the respiratory tract." It also reported the result that "Death rates of white men in Atlanta due to respiratory tract cancer, pneumonia and pulmonary tuberculosis were substantially higher in areas with polluted air than in areas with clean air." This was at a time when no measurements of pollutants were being made, but concerns were beginning to stir. In the 1950s and especially the 1960s, there were a large number of editorials and articles about the unhealthy state of the air in Atlanta.

Although other fuels did not produce as much black smoke as soft coal, there were very few "clean" fuels used to heat homes and factories early in the twentieth century. The wood that was used to heat most Georgia homes in the early 1900s did not burn much cleaner than coal, and it was used in vast quantities. Surveys between 1900 and 1920 showed that five to six million cords of wood per year were used as fuel in the state, about 2 cords per person[11] (a cord is a stack of wood 8 feet long by 4 feet high and 4 feet wide). Eleven percent of that wood was used in the towns and cities where at the time about 20 percent of the people lived. Ninety-five percent of rural farm homes in 1940 were heated with wood.[12]

Recently there have been news articles about the quality of air in homes, and even some assertions that it is more polluted than outside air. Even the USEPA concluded that "Measured indoor air concentrations of PM-2.5 and PM-10 generally exceed outdoor air concentrations (often by a factor of two) except in areas where outdoor concentrations are high (e.g., Steubenville, Ohio and Riverside, California)."[13] PM-2.5 and PM-10 are particles smaller than 2.5 and 10 micrometers diameter, respectively; PM-10 includes most of the airborne particles. These particles are invisible to the human eye. By comparison, most bacterial cells are between 1 and 5 micrometers in diameter. Air in the modern home probably does not come close to the poor quality of that in houses which had open fireplaces, wood-burning heaters and cook-stoves, kerosene lamps, and candles. According to Heinsohn and Kabel (1999), the concentration of carbon monoxide, suspended particles, and several organic gases in homes using wood-burning stoves is generally comparable to the concentration of these materials in public places where smoking is allowed. During cold weather, of course, houses in the old days were closed as much as possible, but in warmer weather they were open so that foul air did not accumulate in them.

Although not as steady as the smoke pouring year-around from homes, factories, and trains, burning of woods and fields must have contributed greatly to air pollution at certain times of the year. It was common practice in the early part of the twentieth century to burn the woods as described in chapter 2. Holbrook (1943), in a book on the effects of fire on American forests, related the following about the situation in Georgia in the 1930s and 1940s.

> During twenty months of 1931 and 1932, the pine forests and hardwood swamps and bays of the coastal region of the Carolinas, Georgia, and the northern half of Florida were honeycombed with fire. The smoke was so thick that travel on the highways and railroads was difficult.
>
> In south Georgia townsmen by the thousand turned out to save their villages.
>
> William Hagenstein, a West Coast forester and careful observer, who visited the Georgia woods in the spring of 1941, said he was struck most forcibly by the unattended fires in the forests, burning on every hand for miles and miles along the roads.

In the 1930s, between four and five million acres of forest were burned in Georgia each year,[14] some accidentally, but most on purpose. According to visible evidence on forest trees in a survey in 1934–1936, fires occurred at irregular intervals on 77 percent of the forest area of the state.[15] This burning served several purposes, such as reducing the growth of undesirable plants, promoting the growth of grasses in the woods for grazing by cattle, and encouraging certain wild game species, especially quail. Stoddard (1963a), a quail expert, noted this additional benefit for burning forest, "The speaker has long noted the favorable effects of frequent fires in reducing the numbers of ticks and chiggers, both in pineland and open ground of the deep Southeast." Whatever the benefits, the burning of four to five million acres must have made the air quality worse for some of the population during part of the year.

Annual burning of the forest was common practice in the early part of this century, but the practice was not limited to forest. Fields that accumulated a growth of broomsedge and other weeds, or crop residue from the previous year were burned to make them easier to cultivate. I remember, as a young boy, seeing my grandfather rake cornstalks into piles and burn them leaving the field clean, even before it was plowed. There was also burning of trash at every homestead. City and community dumps were scattered across Georgia and the refuse carried to these dumps was burned regularly. As late as 1971 there were 416 open dumps in Georgia, and 81 percent of them allowed open burning.[16] Nowadays, there are fewer acres in cultivation and fields and woods are seldom burned. Most cities and counties require a permit to set fires in the open. Trash is no longer burned at home, and modern landfills do not burn trash, but cover it with soil the same day it is dumped.

Dust, however, may have been a more serious and widespread air pollutant in the country than smoke before the 1950s, although it was not considered a pollutant at that time. As a boy I plowed many days in the field, early on behind a mule, and later on a tractor. Dust was a frequent occupational hazard. It was worst at peanut harvest time, when we put the dried plants through a stationary thresher. The dried peanut pods and plants still held much of the soil surrounding the roots even though they had been harvested and "cured" for weeks in "haystacks" in the field. One of the crew of about six would have to stand behind the thresher in the midst of all the dust and dirt and throw the peanut vines (the chaff removed from the peanuts) up on the tray of the hay baler to be pressed into bales. He inhaled and ingested the most dust. At the end of the day we were covered with dust inside and out. There was so much dust in our lungs that our spit was black for a week after finishing the peanut harvesting. Fortunately, days as dusty as those around a peanut thresher were fairly infrequent in the life of most farmers, and no one got sick or died from dust—at least no one that we knew. However, the dusty days in dry, cultivated fields, around peanut harvesters, in cotton gins, and dozens of other agricultural jobs fifty years ago certainly involved breathing air that was over the modern limits for suspended particles.

Anyone who lived close by a dirt road early in the automobile age knew the nuisance, if not the unhealthy conditions, of dust settling on the porch and in the house every time an automobile passed. Jimmy Carter describes the dust and dirt that settled on and inside the house he grew up in 50 feet from a dirt road in Archery, Georgia.[17] It was a common experience. An astounding statistic from the USEPA is that over 352,000 tons or 30 percent of particles smaller than 10 micrometers in diameter (PM-10) released into the air of Georgia in 1994 came from unpaved roads.[18] These emissions numbers must be taken with a grain of salt, however, because most of the dust raised on unpaved roads settles near the road and affects few people. In the early-to-mid-twentieth century, however, many Georgians lived by dirt roads. Eighty-seven percent of Georgia's farms were on dirt roads in 1930.[19] If 352,000 tons of fine dust came from unpaved roads in 1994, then the amount of dust from unpaved roads in the 1940s and 1950s must have been many times higher.

The dust of the countryside was not the reason, but beginning in the mid-1900s, there was an awakening to the dangers of air pollution in the United States, and Georgians became aware of the problem. In addition to the article mentioned above about the association between respiratory illness and air pollution in Atlanta, there were others exhorting action on air pollution. Several pleaded for preventive action so that Atlanta would not become as polluted as some of the more industrialized cities. One editorial in the *Atlanta Journal* (1965) entitled "Our Filthy Society" stated that during a severe pollution episode over the northeastern United States, "Hundreds of thousands, maybe millions of people were within a breath or so of death." Such hyperbole is common in the beginning of a campaign to mobilize public opinion and when the situation is known only from speculation and scattered facts. That was the situation in the 1950s and 1960s and the nation was persuaded that air pollution was serious and needed remedy.

Monitoring the Air

As is the case for most pollution, measurements and descriptions of air pollution focus on the trouble spots and tend to overestimate the general problems. At the beginning of air quality monitoring in Georgia, sampling stations were set up only in the cities where air quality was expected to be worst. Even in the late 1990s assessments were biased toward areas where air quality was worst. The bias is indicated in the Ambient Air Surveillance Report of the Georgia Department of Natural Resources (1996b) as follows; "The number and location of the individual sites vary from year to year, depending on a variety of reasons which include: availability of long term space allocation; citizen complaint; regulatory need; etc." It was further stated that, the general objectives governing selection of sites were "1) to determine the highest concentration expected to occur" and "2) to determine representative concentrations in areas of high population density." Although the fourth objective was "to determine the general background concentration levels," it is clear that the emphasis has been on determining the most polluted conditions in the state. In a report on air quality in Savannah, the Georgia Conservancy stated that nine sampling sites in Chatham County "were located at expected maximum concentration points."[20]

Such an emphasis is understandable because the most severe pollution problems are where the most people live and are the ones needing most attention. However, such criteria for collecting data tend to bias the reports of pollution toward overstatement rather than a balanced account. It should be pointed out that although most sampling stations were in cities and some were intentionally located in the most polluted areas, a high percentage were set up in neighborhoods expected be the most sensitive to pollution, but not likely the most polluted. In 1985 when measuring for airborne particles was at its height, there were fifty-eight stations in Georgia and half were at schools, hospitals, or health centers. Location of monitoring stations at or near such institutions may have tended to lessen the bias toward overestimating pollution.

One of the difficulties in judging the air quality of Georgia in the past is the lack of measurements. Prior to the 1950s there were

Table 5.1. National Ambient Air Quality Standards set by EPA.

Pollutant	Year started	Standard[1]	Time interval
Particulate matter	1971	75 µg/cu. m.	Year
	1971	260 µg/cu. m.	24 hour
PM-10	1987	50 µg/cu. m.	Year
	1987	150 µg/cu. m.	24 hour
PM-2.5	1997	15 µg/cu. m.	Year
	1997	65 µg/cu. m.	24 hour
Sulfur dioxide	1971	140 µl/cu. m.	24 hour
	1971	30 µl/cu. m.	Year
Ozone	1971	80 µl/cu. m.	1 hour (phot. Ox.)[2]
	1979	120 µl/cu. m.	1 hour
	1997	80 µl/cu. m.	8 hour
Nitrogen dioxide	1971	53 µl/cu. m.	Year
Carbon monoxide	1971	35,000 µl/cu. m.	1 hour
	1971	9,000 µl/cu. m.	8 hour
Lead	1978	1.5 µg/cu. m.	Calendar Qtr.

[1] µg/cu. m. = micrograms per cubic meter; µl/cu. m. = microliters per cubic meter
[2] Ozone was first regulated as photo-oxidants.

no measurements of the pollutants in the air. The US Public Health Service made the first systematic measurements of air quality in Atlanta in 1953 at the Fulton County Health Department. The measurements consisted of total particles suspended in the air. Within two or three years, sulfur dioxide measurements were started, and stations were added in Savannah, Macon, Columbus, and Augusta. The stations outside of Atlanta were designated as secondary or tertiary, meaning measurements were made every two or three years rather than annually. In the 1960s several other stations were added to the air-monitoring network, and by 1985, there were fifty-eight stations measuring particles in the air. In that year the Georgia Department of Natural Resources established the Georgia Air Quality Monitoring Network for the measurement of several air pollutants, and since then a more complete description of air quality in the state is available. I have used the data from both the older and more recent records to show that air pollution problems have become smaller, not greater, since the mid-1900s. Values for trends in the early years are subject to fluctuation and some uncertainty because of irregular sampling of the various locations and cruder methods of measurement. For some stations in some years no measurements were made, or were too few to report an average for that year. Relocation of stations within cities probably caused some year-to-year differences in pollutant levels.

With increases in the awareness and documentation of pollution there came legislation to control the levels of pollutants in the air. The national Clean Air Act of 1970 provided for the setting of standards to protect the health of citizens, and standards were set for six "criteria pollutants." Table 5.1 gives the criteria pollutants and the "Primary Standards" established to protect human health. Levels over the values in Table 5.1 for the specified measurement time are considered violations of the standards, although actual violations require averaging over different numbers of days or years depending on the pollutant. Secondary Standards were set equal to or lower than the Primary Standards to protect the public welfare, that is, to protect property such as crops, animals, and buildings. The standards were put in place in 1971, except for lead, which was set in 1978. Some of the standards have changed over time to reflect changes in the methods of measurement or changes in knowledge about health hazards. Where appropriate, pollutant concentrations in Georgia air will be compared to the primary standards.

Dust, Dirt, and Particles

The emphasis on suspended particles in the 1950s and 1960s was probably because they are the most obvious and the easiest to measure. Particles that are the most noticeable in the air are those in smoke and dust. Suspended particles range in size from 0.1 to about 100 micrometers, but most of the particles are less than 10 micrometers in diameter. Particles are suspended in the air because of their small size, and the smaller they are the longer they remain suspended and the farther they move in air currents. Airborne particles may be viewed as a single pollutant, but are really a mixture of sizes of particles from many sources, containing many different chemicals. Suspended particles in the air include dust, smoke, pollen, bits, pieces of organic and mineral materials, and even microscopic particles of chemicals or chemical mixtures. Particles may also be classified as primary and secondary; primary particles are those that enter the air as solids, and secondary particles are those that form in the air from gases or coagulation of smaller particles.

The early references to particles in the air were as Total Suspended Particulates (TSP). Measurement of TSP was typically made by drawing a known volume of air through a filter and weighing the filter before and after drawing the air through it. The difference in weight, divided by the volume of air drawn through the filter is the concentration of particles in the air. The largest particles drawn into the collector are about 100 micrometers in diameter. About 50 percent of particles in industrial areas are mineral and about 25 percent are combustion products.[21] Samples from undeveloped areas have a higher percentage of minerals (about 77 percent) and lower combustion products (about 7 percent).

In some cities in the past, particle concentrations were high particularly during pollution episodes. Some such pollution episodes resulted in death rates considerably above the normal. In the Donora, Pennsylvania smog of 1948, pollutants were not measured. However, about six months later, particle concentrations averaged 2,630 micrograms per cubic meter during a temperature inversion (see description of inversion in section on ozone) and 1,190 micrograms per cubic meter after the inversion. In a famous London smog of 1952, during which there were 4,700 deaths above the normal rate, smoke ("smoke" measurement is the European near-equivalent of TSP) concentrations reached 6,000 micrograms per cubic meter.

The smallest of the suspended particles are the most dangerous because they can penetrate deepest into the lungs and because they are likely to contain the most reactive chemicals. The larger particles are likely to be more benign pollen and nearly inert dust. Of course, some of the pollen particles are very reactive and cause hay fever symptoms, but pollen is not considered a pollutant. In the 1970s the USEPA established a standard for TSP. In recognition of the fact that smaller particles are more harmful to the respiratory system, the USEPA established new standards for the smaller particles. In 1986 a new classification was created and designated PM-10. This classification includes particle sizes below 10 micrometers in diameter, and in 1997, a newer, smaller standard was proposed (PM-2.5), for particles smaller than 2.5 micrometers in diameter.

In the southeast, about 60 to 70 percent of PM-10 is made up of particles smaller than 2.5 micrometers in diameter, that is PM-2.5. Fifty percent or more of the weight of these fine particles is made of sulfate and 30 to 40 percent are organic compounds.[22] Much of the sulfate comes from SO_2 in the air when it reacts with water and the molecules of sulfate coalesce to form particles or are adsorbed onto other particles. On the other hand, the coarser particles (2.5-10 micrometers diameter) were about one-half mineral matter, probably soil particles, and contained only about 6 percent sulfur compounds.

There are differences in the amount of suspended air particles between cities, between urban and rural areas, and even at different locations within cities. Urban stations in the southeastern United States have about 5-10 micrograms per cubic meter more particles than rural stations. It is certain that suspended particles are higher downwind from factory smokestacks than in suburban areas. It is not certain, however, that location of sampling stations are always in areas that are representative of the cities. For example, there were twelve sampling stations in Atlanta in 1976, and the yearly average ranged from 37 micrograms per cubic meter at E. Rivers School to 72 micrograms on Marietta Boulevard. The corresponding twenty-four hour maximum values were 65 and 155 micrograms per cubic meters. Twelve stations may have provided a representative yearly average for Atlanta (58 micrograms per cubic meter), but in the same year only one station reported for Augusta, hardly enough to characterize the air for the whole city.

The concentration of particles in the air was never as high in Georgia as in some of the cities more famous for smog. However, prior to the 1970s particle concentrations in Atlanta were consistently above the standard of 75 micrograms per cubic meter set up by the USEPA in 1971 (Figure 5.1), and frequently were above that level in several other Georgia cities. By the early- to mid-1970s, particulate pollution was already reduced to near or below the USEPA standard. From 1973 to 1977, only Rossville in Walker County, a suburb of Chattanooga, exceeded the standard with an average of 77 micrograms per cubic meter (Table 5.2) and the standard was exceeded for only two of the five years at 87 micrograms per cubic meter. Only one other location, Fulton County, exceeded the standard in one year with a concentration of 83 micrograms per cubic meter. According to Kundell and Dorfman (1994), there were four locations in Georgia listed as nonattainment areas under the Clean Air Act of 1970 (Atlanta, Savannah, Sandersville, and Rossville), but by 1982 all had been removed from the list. The location with the lowest concentration of particles was Brunswick with 45 micrograms per cubic meter (Table 5.2).

The yearly averages do not reflect the worst cases of air pollution in the 1970s. The nature of air pollution is such that hazy, smoky days come and go, depending on the weather and the source of pollutants. Respiratory ailments that are aggravated by smoke and dust are most severe when pollutants are at their maximum.

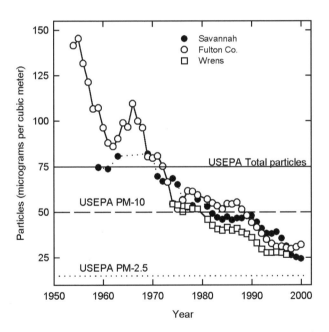

Figure 5.1. Decreases in the yearly average of suspended particles in the air of Fulton County, Savannah, and Wrens, Georgia during the last half of the twentieth century. All stations for Fulton County, ranging from 1 to 14, were averaged. Horizontal lines represent the USEPA standards set in 1971 for total suspended particles, in 1987 for PM-10, and 1997 for PM-2.5. From US Department of Health, Education and Welfare (1958, 1962), the US Environmental Protection Agency (1973–1998, 1994–2001), and the Georgia Department of Natural Resources (1985–2000).

Maximum twenty-four-hour values in Table 5.2 show that concentrations in the mid-1970s could be six or seven times as high as annual averages. What's more, the locations with the highest yearly averages did not necessarily have the worst pollution days. Savannah (Chatham County) had the highest twenty-four-hour values in the mid-1970s averaging 455 micrograms per cubic meter. Albany, in Dougherty County, also had high pollution days during that period with average maximum twenty-four-hour values of 414 micrograms per cubic meter.

Yearly levels of suspended particles in the mid-1950s in Atlanta were between 140 and 150 micrograms per cubic meter (Figure 5.1). By the mid-1990s average yearly levels for Atlanta had

Table 5.2. Maximum and average concentrations of suspended particles for the years 1973-1977 and 1992-96. From the U.S. Environmental Protection Agency (1973-1998, 1994-2001) and the Georgia Department of Natural Resources (1985-2000).

Location	Suspended particles					
	------24 hr maximum[1]------		Percent reduction	------Annual average------		Percent reduction
	1973-77	1992-96		1973-77	1992-96	
	µg. per cu. m.[2]			µg. per cu. m.		
Rossville[3]	240	56	77	77	33	77
Rome	164	70	57	59	25	58
Dekalb County	138	61	56	–	28	–
Fulton County	371	68	82	58	35	40
Augusta	153	62	59	56	28	50
Macon	205	67	67	53	34	36
Savannah	455	66	85	61	39	36
Albany	414	58	86	54	29	46
Brunswick	173	74	57	45	33	27
Average	257	65	75	58[4]	32[4]	45[4]

[1] These values are averages of highest concentrations each year during the period. [2] micrograms per cubic meter [3] In Walker County near Chattanooga, TN. [4] These averages do not include Dekalb County.

dropped to about 35 micrograms per cubic meter. The downward trends were similar in other Georgia cities. The Richmond County Department of Health (1964) evaluated the quality of air in Augusta in the early 1960s and concluded that "Consequently, suspended particulate matter, although significant, is apparently not a serious air pollution problem in Augusta." This was based on a three-year average of 78 micrograms per cubic meter, compared to a 117-microgram-per-cubic-meter average for eighty-seven other United States locations. In the 1990s, yearly values for Augusta averaged 28 micrograms per cubic meter (Table 5.2). Rossville in Walker County was among the top locations for suspended particles in the mid-1970s (average of 77 micrograms per cubic meter), but had a yearly average of only 33 micrograms per cubic meter in the mid-1990s. So, if the situation in Augusta in the early 1960s could be judged "not a serious air pollution problem," today all of Georgia air can be declared much cleaner with much stronger justification.

It is not just the industrial areas, whose air was the foulest in the mid-1900s, that have become cleaner. Although the larger metropolitan areas had higher concentrations of particles in the air near mid-century, all areas have reduced their worst pollution days by more than 50 percent and annual averages by 27 percent or more (Table 5.2). The locations with the smallest percentage of improvement also had the lowest particle levels in the early 1970s. Wrens, a small town about 30 miles southwest of Augusta, had particle levels between 50 and 60 micrograms per cubic meter in the mid-1970s. In the 1990s they were half that amount (Figure 5.1). Because there was never much industrial development around Wrens, it is unlikely that the reduction is due to cleanup of factory smokestacks. More likely it resulted from less use of wood for heating, less open burning, and decreased dust from cultivated fields. The acreage of row crops in that county (Jefferson) has decreased from about 130,000 in 1945 to 63,000 in 1997. There were probably other contributing factors such as the cleaning of factory and power station smokestacks in towns and cities at distant locations, and the paving of roads, not only near Wrens, but all across Georgia.

In the 1990s, all locations were not only below the older TSP standard, but also below the newer PM-10 standard. Actually, measured amounts of PM-10 are not much lower than total suspended particles and the USEPA allowed Georgia to use TSP as a surrogate for PM-10. Therefore, only a few stations measured PM-10 in Georgia before the late 1990s. In Fulton County, PM-10 has been measured at three stations during some years since 1989.[23] The average of those measurements was 78 percent as high as the average of total particles at other stations in the county. So some, but not all, of the decrease in the values for Savannah and Fulton County during the 1990s is due to the changeover from measuring total particles to PM-10.

The pleasant decrease in air particles has not stopped the gloomy descriptions of Georgia's air. As recently as May 2000 scientists at a conference at Stone Mountain were proclaiming the danger of the particle pollution in Georgia. Dr. Howard Frumpkin, chairman of Emory University's Department of Environmental and Occupational Health is quoted as saying, "A 1989 study showed particulate matter—especially the tiniest variety, 30 times smaller than a human hair—may be responsible for about 950 deaths in metro Atlanta each year." At the same conference, Dr. Glen Cass, chairman of Georgia Tech's School of Earth and Atmospheric Sciences is credited with the following assessment from a study of Atlanta's air: "The study suggests Atlanta's air may become more polluted than Los Angeles' if present trends continue. In Atlanta, the air appears to be getting dirtier…."[24] If the numbers taken from government documents and shown in Figure 5.1 are even close to correct, Atlanta's air is certainly much cleaner.

Along with the cleaning of the air have come ever-tighter restrictions on pollution levels and increased emphasis on smaller particles. By the mid-1970s, air particles in Atlanta had been reduced from nearly 150 to less than the 75-microgram-per-cubic-meter standard, and in the 1990s they were below the new PM-10 standard (Figure 5.1). In 1997, USEPA proposed a further restriction on particle pollution by setting a standard limit for particles smaller than 2.5 micrometers (PM-2.5) at 15 micrograms per cubic meter. The Georgia Department of Natural Resources (2001c) tested the air at twenty-seven locations for nine months in 2000 and found twenty-three of the locations exceeded the PM-2.5 standard. So in spite of the remarkable cleaning of the air depicted in Figure 5.1, much of Georgia may be in violation of the newest USEPA particle pollution standard.

Sulfur Dioxide and Acid Rain

The second major pollutant to be monitored in Georgia air was sulfur dioxide. It had long been known as a major pollutant from burning of coal and when acid rain became an environmental concern it was widely blamed on sulfur dioxide. Sulfur dioxide also comes from natural sources, such as volcanoes; worldwide about 50 to 75 percent of it. But in eastern North America, it has been estimated that man-made sources account for more than 90 percent of sulfur dioxide and nitrogen oxides.[25] According to the USEPA, 70 percent of sulfur dioxide pollution in the United States in 1980 was from burning of coal, but other fuels also contribute.[26] The USEPA also estimated that 75 to 80 percent of sulfur dioxide emissions in Georgia during the early 1970s were from burning coal.[27]

In a 1991 assessment of air quality in Georgia, Kundell and Swanson attributed a decrease in visibility in Georgia skies to an increase in sulfur dioxide. They stated that, "Since the 1950s a decline in summer visibility in Georgia of 50 to 100 percent is strongly associated with the increase in sulfur emissions during that time." A similar assessment of the Southeast was made by the National Research Council (1986): "From data on SO_2 emissions, reduction in visibility, and sulfate in Bench-mark streams since about 1970, we conclude that the southeastern United States has experienced the greatest rates of increase in parameters related to acid deposition." As will be seen in a later section, visibility in Georgia skies has improved since the 1970s along with decreases of sulfur dioxide emissions and levels in the air. The development of electricity and natural gas for heating eliminated hundreds of thousands of home fires in Georgia fueled by coal and wood. The simple substitution of electricity for coal may not have resulted in less sulfur dioxide pollution, because the power plants used coal to generate electricity. However, the release of contaminants was far removed from most homes and much easier to regulate by government action. Later, switching to gas-fired power plants, governmental restrictions on the use of high-sulfur fuels, and requirements for scrubbing of sulfur dioxide from smokestacks resulted in less air pollution.

Changes in sulfur dioxide pollution in Georgia and the United States have followed increased and decreased use of coal during the twentieth century. The USEPA has estimated the trend in emissions in the United States since 1900 based mainly on fuel usage.[28] For Georgia, sulfur dioxide pollution was estimated at about 100,000 tons in 1900 and remained near or below 200,000 tons from the 1920s to the 1960s. From 1960 to the early 1980s, emissions rose from 200,000 to just over 1,000,000 tons (Figure 5.2B). Since then, however, they have decreased about 40 percent, with estimates in 1997 being 639,000 tons. Likewise, USEPA estimates that sulfur dioxide emissions in the United States have decreased by 35 percent from 31 million tons in 1970 to 20 million tons in 1997. By these estimates, sulfur dioxide emissions in the US are lower now than at any time since the 1920s, and in Georgia, emissions in the late 1990s were at mid-1970s levels.[29]

With such large decreases in estimated emissions of sulfur dioxide a decrease in air levels should be expected. In Fulton County where longest-running measurements have been made, the yearly average sulfur dioxide concentrations have dropped from about 15 microliters per cubic meter (parts per billion) in the mid 1970s to about 3.5 microliters per cubic meter in 2000, a decrease of 75 percent. (Figure 5.2A). The line in Figure 5.2A represents all of the measurements in Fulton County and the decrease corresponds closely to the decrease in average levels for the United States. Large decreases in annual average sulfur dioxide occurred all across Georgia from 1985 to 2000, ranging from 52 percent at

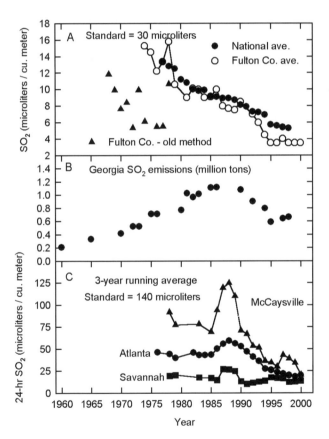

Figure 5.2. Annual average sulfur dioxide concentrations in Fulton County and the nation (A), millions of tons of sulfur dioxide emitted in Georgia (B), and maximum 24-hour sulfur dioxide levels at three Georgia locations (C). From the US Environmental Protection Agency (1973–1998, 1994–2001) (concentrations and emissions), the Georgia Department of Natural Resources (1985–2000) (concentrations), and Gschwandtner et al. (1985) (emissions).

Milledgeville in the center of the state to 77 percent at McCaysville near the Tennessee border (Table 5.3). The decreases since the 1960s and 1970s are actually larger than shown in Table 5.3, because as shown for Fulton County in Figure 5.2A, concentrations had already dropped by one-half from 1974 to 1985. Records for some other locations are also available for the late 1970s. At Rome, for example, sulfur dioxide dropped from about 15 microliters per liter in the late 1970s to 7.4 microliters in 1985, and at Putney in Dougherty County near Albany the level dropped from 9.2 microliters per liter in 1978 to 5.9 microliters in 1985. Measurements by an older, less reliable method indicated a downward trend for Fulton County even earlier, from 1968 to 1977, as shown by the tri-

Table 5.3. Averages of annual sulfur dioxide concentrations and percent decrease in Georgia air. From the U.S. Environmental Protection Agency (1973-1998, 1994-2001) and the Georgia Department of Natural Resources (1985-2000).

Location	Years	Average conc. μl/cu. m.[1]	Percent decrease
Sampled 10 years or more			
McCaysville (Tenn. border)	1985-2000	9.4	77
Stilesboro (Cartersville)	1985-2000	3.4	62
Rome	1985-2000	4.2	65
Atlanta (Fulton Co.)	1985-2000	6.5	61
Savannah	1985-2000	3.4	53
Sampled less than 10 years			
Milledgeville (6)[2]	1986-2000	3.7	52
Augusta (6)	1985-1998	3.6	53
Macon (7)	1985-2000	3.4	63
Albany (6)	1985-1998	3.3	83

[1] microliters per cubic meter [2] Number of years measured.

angles in Figure 5.2A. In this older method air was bubbled through a solution that trapped the sulfur dioxide and it was later found to underestimate levels of the gas. But the older method may show the correct trend, because in 1978 it indicated an increase to nearly the 1968 level, a year when the newer method also showed increased sulfur dioxide.

The early decreases in sulfur dioxide in the air do not coincide with trends in emissions. These calculated emission trends increase until about 1985 before declining (Figure 5.2B). This difference in patterns of sulfur dioxide emissions and concentrations is even more extreme than that observed by Darlington et al (1997) for the United States. They suggested that downward changes in emissions were being underestimated on a national basis.

In addition to yearly averages, the highest twenty-four-hour average sulfur dioxide levels have been reported since the late 1970s. These twenty-four-hour maximums are two to five times higher than the yearly average of all measurements (compare lines for Atlanta in Figure 5.2A and C). The yearly maximum values rose to a peak in the late 1980s and since then have dropped by 50 percent for locations in North Georgia as illustrated in Figure 5.2C for Atlanta and McCaysville. Maximum sulfur dioxide levels did not decrease much at Savannah because they were never very high, but concentrations in Atlanta area dropped to near those in Savannah in the mid-1990s (Figure 5.2C). The low sulfur dioxide levels for Chatham County in the late 1970s are significant because The Georgia Conservancy (1979) reported that Chatham County was the most industrialized area in Georgia in the late 1970s with approximately one-third of the 350 major stationary pollution sources in the state being located there. Apparently it was not the kind of industry that generated large amounts of sulfur dioxide.

Because sulfur dioxide reacts with water in the air to form sulfuric acid, it has been blamed for acid rain that was such an environmental watchword for about a decade in the 1970s and

1980s. At that time some lakes in Europe, eastern Canada, and the northeastern United States were becoming more acid than in the past and some had declining fish populations.[30] Trees at high elevations, including some in the Smoky Mountains, exhibited discolored leaves, lost leaves, and died, possibly because of acidity of the fog that frequently hangs over the peaks. Corrosion and deterioration of buildings, sculpture, and exposed valuables around the world were blamed on acid rain. An example of the concern expressed for the southeastern United States is given in the National Clean Air Coalition and Friends of the Earth Foundation 1984 report in which Olsen et al state that "In 1955, most areas of the South received rainfall of average pH, about 5.6. By 1979, however, all states in the South were receiving rainfall with an average annual pH of 4.6 or less, ten times more acidic than in 1955–56." The "ten times" estimate in the assessment comes from the fact that a drop of one pH unit represents a ten-fold increase in acidity. In an assessment of air quality in Georgia, Kundell and Swanson (1991) projected the possible damage by acid rain. One example of projected damage was, "Acid rain and ozone could also damage materials including marble and limestone, paints, and zinc and steel. If acid rain-damaged materials are repaired or replaced, those costs could reach millions each year statewide." In a statement similar to that of Olsen et al,[31] Kundell and Swanson also stated that, "Average measured pH of rain in Georgia is 4.5, more acidic than natural rain."

Actually 4.5 is about the lower limit for pH of rainwater in Georgia. Only four of the twenty-four averages in Table 5.4 were that low. For three of these (Athens 1976–1977, Okefenokee Swamp 1974–1975, and Brier Creek) pH was measured for only one or two years, and for the fourth, Summerville in Northwest Georgia, there has been an increase in pH from 4.3 in 1985 to 4.7 in 2000. Measurements made at Athens over a ten-year period (1989–1998) averaged 4.6. The pH of rain is not the same all over the state, but as is shown in Table 5.4 values are higher (less acid) in southern Georgia than in the north. The average value for seven

Table 5.4. Average pH of rainwater at locations in Georgia for the years of record.

Location	Years of measurement	Average pH	pH[1] change	Source
Hiawassee	1985-2000	4.6	+0.15	GDNR[2]
Blairsville	1974	5.5	—	Giddens, 1975
Dawsonville	1985-2000	4.6	+0.13	GDNR
Summerville	1985-2000	4.5	+0.35	GDNR
Calhoun	1974	5.0	—	Giddens, 1975
Athens	1989-98	4.6	none	GDNR
Athens	1974	4.7	—	Giddens, 1975
Athens	1976-77 (12 Mo.)	4.2	—	Haines, 1979
Eatonton	1986-98	4.6	+0.25	GDNR
Griffin	1978-2000	4.6	none	GDNR, NADP[3]
Henry County	1974	4.8	—	Giddens, 1975
Brier Creek	1985-86 (12 Mo.)	4.4	—	Buell and Peters, 1988
McDuffie County	1985-2000	4.7	none	GDNR
Midville	1974	5.1	—	Giddens, 1975
Plains	1974	5.0	—	Giddens, 1975
Bellville	1983-2000	4.7	none	GDNR, NADP
Tifton	1974	4.8	—	Giddens, 1975
Tifton	1983-2000	4.8	none	GDNR, NADP
Waycross	1985-92	4.8	—	GDNR
Okefenokee Swamp	1974-75	4.5	—	Rykiel, 1977
Okefenokee Swamp	1997-2000	4.6	—	NADP
Buena Vista	1992-95	4.8	—	Huntington, 1996
Lumpkin	1992-95	4.9	—	Huntington, 1996
Glynn County	1985-88	4.6	—	GDNR

[1] pH change was tested only at those locations with at least 10 years of measurements. [2] Georgia Department of Natural Resources, 1985-2000. [3] National Atmospheric Deposition Program / National Trends Network, 2001.

locations north of the fall line, which were sampled for ten years or more, was 4.6. Two locations in South Georgia sampled ten years or more had an average pH of 4.75. Shorter-term measurements also tended to be higher in southern Georgia, except that a three-year average for Glynn County was 4.6 and a one-year average for three locations in the Okefenokee Swamp (1974–1975) was 4.5.[32] However, 1997–2000 measurements at the swamp averaged 4.6 and at Waycross, only a few miles away, the 1985–1992 average was 4.8, similar to that of other South Georgia locations.

If rain became more acid at some earlier time in Georgia, it has not grown worse since the first measurements were made, and in at least four locations rain has become significantly less acid. The longest running pH record was begun in 1978 near Griffin by the University of Georgia Agricultural Experiment Station. At that location, the pH of rainwater is unchanged over twenty-three years, remaining at about 4.6 (Figure 5.3A, Table 5.4). In contrast to my analysis, Walker and Melin (1998) concluded that pH increased at this location from 4.43 in 1984 to 4.79 in 1996. An analysis of measurements from three other locations (Tifton, Bellville, and McDuffie County) beginning in 1985 or 1986 showed no change (Table 5.4), but at Dawsonville, Summerville, Hiawassee, and Eatonton pH actually increased by 0.1–0.3 units. When the yearly values are averaged across all eight of these locations to minimize the year-to-year variation, pH increased from 4.6 in 1986 to 4.7 in 1998. So if there has been a change in acid rain since the mid-1980s, it is a change toward less acid in spite of a statement as late as 1994 by Kundell and Dorfman that the pH of rain in Georgia was 4.5.

Acidity of rainwater is a result of the balance between anions, such as sulfate (SO_4^{-2}) and nitrate (NO_3^{-1}), which lower the pH and cations, such as calcium (Ca^{+2}), and sodium (Na^{+1}), which raise the pH. If there are equal numbers of anions and cations of equal strength in solution, the pH is 7.0 or neutral. If there are not enough cations to balance the anions, then the balance is made up by hydrogen ions from the water. The pH is a reciprocal measure of the concentration of hydrogen ions, so an increase in hydrogen ions lowers the pH. One of the main pollutants that causes acid rain is sulfur dioxide (SO_2) which is converted to sulfate in rainwater. If there are not cations to balance the sulfate, then the hydrogen ion from water combines with sulfate to form sulfuric acid (H_2SO_4). Nitrogen oxides in the air can be converted to nitric acid (HNO_3). Even carbon dioxide (CO_2) in the air can be converted to a weak acid, known as carbonic acid (HCO_3).

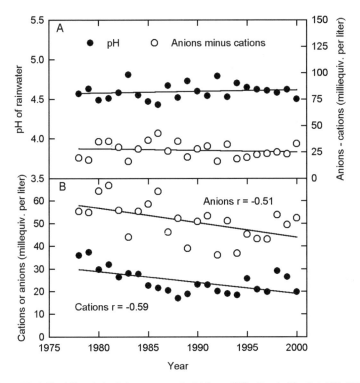

Figure 5.3. Acidity (pH) and chemical components of rainfall near Griffin, Georgia (Pike Co.) 1978–2000. Concentrations have been weighted for rainfall amounts. From records of the National Atmospheric Deposition Program/National Trends Network (2001).

So the question is raised if sulfur dioxide in the air has decreased as shown above, why has acidity of rainwater at all Georgia locations not decreased? The National Academy of Science concluded that "if emissions of sulfur dioxide and nitrous oxides are reduced, there should be an approximately corresponding reduction in acid rain."[33] Acidity may be unchanged even when sulfur dioxide and other acidic pollutants decrease if the basic pollutants that counterbalance the acidic ones also decrease.

Analysis of rainwater collected in Pike County by the Georgia Experiment Station at Griffin shows that both anions and cations have decreased since the measurements began in 1978 (Figure 5.3B). The anions and cations are shown in "equivalent" rather than weight concentrations because it better expresses their pH balancing properties. The capacity of an element to affect pH is determined by its electrochemical charge rather than its weight. The equivalents of the major anions (SO_4^{-2}, NO_3^{-1}, and Cl^{-1}) have decreased about 30 percent since 1978 at Griffin. Although the cation (calcium, magnesium, sodium, potassium, and ammonium ions) levels in rainwater were lower than anions, the levels at Griffin decreased in parallel with the anions. This trend is also shown in Figure 5.3A where the difference in concentration between anions and cations is plotted along with pH. Neither pH nor the difference between anions and cations changed from 1978 to 1999. Lynch et al (1995) found that sulfate concentrations in rain decreased at six of nine southeastern US stations from 1980 to 1992, and calcium and magnesium decreased at eight of the nine. There was no change in pH at six of the nine stations, and at the other three, pH increased. So the trend in Georgia and the Southeast is certainly not toward more acid rain.

Because the pH of rain at Griffin has not changed since 1978 in spite of decreases of acidic pollutants, some doubt must be cast on the degree to which rainfall became more acid from the early part of the century until the 1970s. Increasing acidity of rainfall from the 1950s to the 1970s was based mostly on extrapolations of pH in the 1950s from measurements made at a few locations in the eastern and southern United States, but not in Georgia. The estimates for the South were near 5.6, the pH of water in equilibrium with CO_2 in the air.[34] However, Stensland and Semonin (1982) published "another interpretation" of the pH trend in the United States in which they concluded that the pH values of the 1950s were higher than normal. They concluded that high levels of basic elements in rain in the 1950s, probably resulting from extreme drought and extensive dust storms, caused the high pH values. After adjusting for high levels of calcium and magnesium they calculated the pH values for Georgia in the 1950s to be about 4.6, slightly lower than the average of all the pH measurements in Table 5.4.

Higher pH values for South Georgia (Table 5.4) must result from lower sulfur dioxide and other acid-forming chemicals in air over the southern part of the state. It is known that the most acid precipitation occurs in the eastern mid-west and the northeastern United States and that pH of rain increases progressively further south. The more acid rainfall in the mid-west and northeast is usually explained by the higher concentration of factories and other coal burning facilities in those regions. Not only do the mid-western and northeastern states burn more coal, but the coal mined and used in those regions has the highest sulfur of any in the country. In 1985, four states, Ohio, Indiana, Illinois, and Pennsylvania, burned about one-third of the coal consumed in the United States. For three of those states, Indiana, Ohio, and Pennsylvania, combined sulfur dioxide and nitrogen oxide emissions in 1980 were estimated to be from 66 to 72 tons per square mile, compared to about 18 tons per square mile for Georgia and Alabama.[35]

The concentration of sulfur dioxide decreases from north to south in Georgia in the same way as pH rises, as shown in Table 5.3. Locations near or below the Fall Line that separates North and South Georgia had lower sulfur dioxide than those above, except that levels at Stilesboro, near Cartersville, were similar to those at southern locations. The sixteen-year average for maximum sulfur dioxide concentrations in Atlanta was nearly twice as high as in Savannah and that at McCaysville, just across the state line from Copper Hill, Tennessee was almost three times as high. Copper Hill was infamous in the 1800s and early 1900s for so much sulfur pollution resulting from copper mining that over 10.5 square miles no trees or other vegetation would grow.[36] As late as 1990, a plant across the state line caused damage to vegetation around McCaysville by releases of sulfur dioxide.[37]

Another way to look at the changes in acid rain is to measure sulfur deposited on the land by rain. If the concentration of sulfur in rainwater and the amount of rain are measured, then sulfur deposited per acre can be calculated. Fortunately some such measurements were made as far back as 1941 at the Georgia Experiment Station at Griffin to determine the amount of plant nutrients deposited in rainfall.[38] When those measurements are placed on a graph with more recent ones, it is clear that less sulfur is deposited in the 1980s and 1990s than in earlier decades (Figure 5.4). Although some numbers in the 1980s were as high as in the 1940s and 1950s, the average for the 1990s was 30 percent lower. At this site, the average sulfur that came down in rain was 6.47 pounds per acre per year in the five years measured before 1980, 5.76 in the 1980s and 5.05 in the 1990s.

Sulfur contained in rain is not all of the sulfur that falls from the sky. There is likely to be continuous deposition of sulfur as long as it is being put into the air. Rain likely washes out most of the sulfur in the air at the time of the rainfall, but between rains sulfur and other pollutants come down as "dry deposition." Dry deposition has not been measured in many locations nor for a long period of time, but in some cases it accounts for as much of the acid-forming pollutants as rainfall does.[39] More than 40 percent of the sulfur

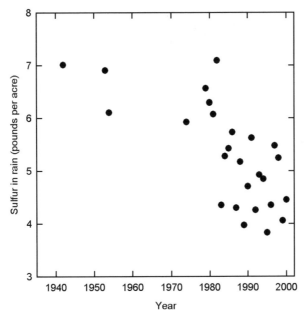

Figure 5.4. The amount of sulfur deposited in rainfall near Griffin, Georgia from 1941 to 2000. From Giddens (1975), Walker (1982), and the National Atmospheric Deposition Program / National Trends Network (2001).

deposited in forests at Panola Mountain southeast of Atlanta from 1987 through 1989 was in the form of "dry deposition."[40]

One of the concerns frequently expressed in the early days of the acid rain controversy was that streams would become acidified and damage the stream ecology. The US Geologic Survey established a Bench Mark Network of water quality stations in the 1960s to study trends of water quality in rural areas unlikely to be substantially polluted. The only station in Georgia was on Falling Creek near Juliette north of Macon. Analyses of water chemistry were first done in 1967. Assessment of the trends in pH and chemical concentrations in the bench-mark streams up to 1981 showed no change in Falling Creek.[41] However, pH measurements up through 1992 show that this stream has actually become less acid since 1967 (Figure 5.5). Acidity has decreased in spite of an increase in the acid-forming sulfate and chloride ions, apparently because the cations (calcium, magnesium, sodium, and potassium) increased

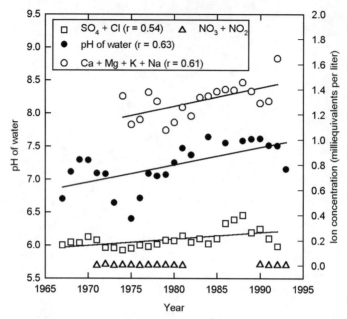

Figure 5.5. The acidity (pH) and ion concentrations of water in Falling Creek near Juliette, Georgia from 1966 to 1992. R-values are correlation coefficients that indicate significant changes in chemistry of the water. From US Geological Survey Records.

more. It is not known what increased the pH of Falling Creek, but it was probably changes in land use or developments in the watershed rather than changes in acid rain.

Actually the most concern for acidification of streams was expressed for those in the mountains of Georgia, which were considered more sensitive to acid rain. Kundell and Swanson, (1991) mention "evidence that stream pH and alkalinity have declined since 1965 in the Blue Ridge Province." Analyses of three locations in northern Georgia where pH measurements were made as far back as the mid-1960s, the Tallulah River, and the Chattahoochee River at Leaf and at Cornelia, show no change in pH with nearly all annual values being between 6.5 and 7.5. A study of fish in two "acid-sensitive streams" of North Georgia in 1983–1984 concluded that "the species of fish commonly found in high elevation headwater streams in the Southern Blue Ridge Province appear to have remained the same during the last 50 years."[42] The same study showed no relationship between fish abundance and pH in twelve streams in the Southern Blue Ridge Mountains. It appears, then, that acid rain has caused no acidification of Georgia streams, and no change in the aquatic creatures that live in them.

Acidity of rainfall may also be affected by oxides of nitrogen that come from the burning of fossil fuels. Nitrogen oxides can be converted to nitric acid in rainwater. According to USEPA estimates, nitrogen oxide emissions in Georgia in the mid-1970s, were about 80 percent as high as those of sulfur dioxide. However, because sulfur dioxide pollution has been more successfully controlled, emissions of nitrogen oxides which come mostly from cars exceeded those of sulfur dioxide in 1997; 691,000 compared to 639,000 tons. Because nitrogen oxides have a controlling effect on ozone, concentrations of this pollutant will be discussed in the next section.

Ozone

The air quality problem that has received the most attention in recent years is ozone in metropolitan Atlanta. It is the topic of most news articles about air pollution in the 1990s, mostly because it frequently rises above the safe level set by the USEPA (Table 5.1). The frequent violation of federal guidelines has endangered allocation of federal funds for highway construction in the Atlanta area. The USEPA issues "ozone alert" days when the level is expected to approach the government standard. On days when ozone alerts are issued, there are increased efforts to convince commuters to carpool or travel by other means than automobiles to reduce emissions of pollutants that cause ozone.

Ozone is the one major pollutant that is not put into the air from man-made sources. Rather, it is formed by the chemical reaction of other pollutants. Two classes of pollutants known as volatile organic compounds (VOC) and nitrogen oxides (NOx) combine under conditions of sunlight, high temperatures, and stagnant air to form ozone. Nitrous oxide (N_2O), also known as "laughing gas," is the main component of NOx, but VOC are a variable mixture of synthetic and natural compounds. It has been estimated that half or more of the VOC in metropolitan Atlanta may come from trees.[43]

Ozone is a simple compound made up of three atoms of oxygen, one more atom than is in the gaseous oxygen molecules that humans need to survive. It is invisible, even though it has become synonymous with smog in recent years. Smog is a word born of the lack of visibility, formed by the combination of smoke and fog. The sound of it is bad enough, however, to have it describe the worst air pollution problem, whatever the cause. The extra oxygen atom in ozone makes the molecule unstable and very reactive. Its ready reactivity makes it useful as a bleaching and sterilizing agent but also makes it a toxic chemical in the lungs. Ozone has a very penetrating odor that is frequently obvious when discharges occur around electrical equipment. Otherwise, the concentration is usually too low in the air for the smell to be noticeable.

"Bad air days" are caused by a combination of pollution and weather. As long as there is enough air movement to sweep away the pollutants, there are few complaints about smog. If there is frequent rain, pollutants are washed out of the air. However, on clear days when wind is light and there is little upward movement of air, pollutants accumulate to high concentrations near the ground causing "smog" and respiratory problems. The normal profile of the air is lower temperature with increased height. When air near the ground is warmed by heat from the sun it rises through the cooler air above. Rising air carries the pollutants with it. When there is a temperature inversion, that is, a warmer air layer overlaying the cooler air, upward diffusion of pollutants is restricted. Since cool air cannot rise through a warm layer, the warm layer acts as a lid on the polluted air. Inversions occur when the wind is slight and especially during weather patterns termed "anticyclones"; patterns associated with stagnant air. Temperature inversions combined with the hot, humid, and sunny conditions of summer result in high levels of ozone if the required pollutants accumulate.

There are usually characteristic daily patterns of ozone buildup. When traffic increases in the city and NOx and VOC accumulate, ozone begins to form and concentrations increase. If the air is stagnant so that pollutants cannot dissipate and if the sun is bright, ozone may concentrate to levels above the USEPA standard (120 microliters per cubic meter) as it did for about five hours on 19 August 1999 on Confederate Avenue in downtown Atlanta (Figure 5.6A). Note that wind speed varied only from zero to 7 miles per

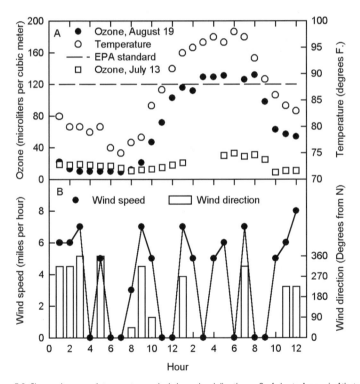

Figure 5.6. Changes in ozone, air temperature, and wind speed and direction on Confederate Avenue in Atlanta on 13 July and 19 August 1999. Only ozone is shown for 13 July. From the Georgia Department of Natural Resources (1985–2000).

hour and wind direction varied from nearly north (0 or 360 degrees) to east (90 degrees) to nearly south, and west (270 degrees). If conditions do not favor buildup of pollutants or the formation of ozone, then ozone remains at low levels, as shown for 13 July 1999. Ozone usually drops to low levels at night. It is obvious in Figure 5.6A that ozone does not start increasing as soon as the sun rises. On both 13 July and 19 August 1999 it did not increase much until 9:00 or 10:00 a.m., somewhat later than the morning rush hour. Highest concentrations occurred in mid to late afternoon. Although the two curves shown in Figure 5.6A are fairly representative of summer days with low and high ozone, the patterns change a great deal depending on cloudiness, wind, temperature, and rainfall.

The nature of the formation of ozone means that the concentration is likely to be highest where the most NOx and VOC pollutants are. Metropolitan Atlanta has the highest ozone concentrations in Georgia (Table 5.5). Out of ten counties where ozone was measured from 1995 through 2000, four of the five with average one-hour maximum concentrations over 120 microliters per cubic meter were metropolitan Atlanta counties, Fulton, Dekalb, Gwinnett, and Rockdale. The same four counties were the only ones that averaged one day or more per year with ozone over the 120-microliters-per-cubic-meter USEPA standard. Fannin County on the Tennessee border and Chatham and Glynn Counties in Southeast Georgia had no days over the standard from 1995 to 2000.

High concentrations of ozone appear to depend not only on slow movement of the air, but on shifting of the air masses. Mulholland et al (1998) found that ozone was highest in downtown Atlanta when air movement was slow, but highest in downwind counties when air movement increased. Chameides and coworkers at Georgia Tech have described a "plume recirculation" that accounts for the majority of the high ozone days that occur in Atlanta.[44] When pollutants build up in the air, ozone begins to form. However, if there is enough wind to blow the pollutants away from the area, neither the primary pollutants nor ozone accumulate to very high levels. If the wind is light and variable as shown in Figure 5.6B, then the pollutants may move only a short distance from the source. When the wind shifts, on the same day or even the next day, it may blow the pollutants back toward the area they came from, "recooking" them and reinforcing the formation of ozone. It is such conditions of "plume recirculation" that causes ozone to accumulate to the highest levels.[45]

The areas of ozone formation and zones of highest concentrations are changeable in the metropolitan area depending on wind speed and direction. The shifting of air masses means that ozone may be under the USEPA standard at a downtown station but above it at Georgia Perimeter College in Decatur, which is about 10 or 12 miles away. If there is a shift in wind direction, the relative concentrations may be reversed the next day. As a result, for example,

Table 5.5. The average 1-hour maximum ozone concentrations and number of days per year exceeding 120 microliters per cubic meter for 10 locations across Georgia during the six-year period, 1995 to 2000. Atlanta metropolitan counties indicated by (M). From the U.S. Environmental Protection Agency (1973-1998, 1994-2001) and the Georgia Department of Natural Resources (1985-2000).

Location	Ave. 1-hr. maximum microliters/cu.m.	No. days per year above 120 microliters/cu. m.
Fannin County	96	0.0
Dawson County	114	0.2
Fulton County (M)	153	7.7
Gwinnett County (M)	133	1.5
Dekalb County (M)	146	4.2
Rockdale County (M)	150	6.3
Richmond County	121	0.3
Muscogee County	110	0.1
Chatham County	99	0.0
Glynn County	96	0.0

Georgia Perimeter College may have more or fewer days over the USEPA standard than the monastery at Conyers east of Atlanta depending on the year. The peak daily ozone level for a twenty-county area around Atlanta was observed by Mulholland et al (1998) to occur at seven separate stations during a twelve-day "ozone episode" in mid-July, 1995. There is a limit, however, to the variation. If ozone is over the USEPA standard anywhere in the Atlanta area, daytime concentrations at other locations are never as low as on 13 July 1999 (Figure 5.6A).

The good news about ozone is that it has not become worse since the mid-1970s when measurements began. Although there is the perception that ozone levels and the number of days exceeding USEPA limits are increasing, Figure 5.7A shows that there has been no change in maximum concentrations (highest one-hour of the year) since the 1970s. These measurements were made at Georgia Perimeter College and the Monastery at Conyers, the longest-running measurements that have been made. Maximum ozone concentrations in Columbus are about three-quarters as high as in metropolitan Atlanta; in fact the 120-microliter-per-liter standard provides a nearly complete separation of the maximum values for the two cities (Figure 5.7A). The maximum levels for Columbus also appear not to have changed since measurements began in 1978. The lack of a trend in ozone concentration agrees with the USEPA conclusion that second-highest hour ozone concentrations in Atlanta, Columbus, and Augusta are also unchanged since 1986.[46]

Ozone alert days have become more numerous in the Atlanta area in the 1990s, but the number of days above 120 microliters per cubic meter has not. It is clear from Figure 5.7B that the number of days above the USEPA standard is cyclic, varying by several days from one year to the next. The variation at Georgia Perimeter College was from zero to fifteen days. In addition to the days above the standard at Perimeter College, Figure 5.7B shows the number of days above the standard averaged across all stations in the Atlanta metropolitan area. Neither the trend at Perimeter College nor the average for all stations shows an increase with time. In fact, the average for stations across the Atlanta area dropped by half, from 5.7 days above the 120-microliter-per-cubic meter standard per year in the 1980s to 2.8 in the 1990s. A 1998 report by the Georgia Department of Natural Resources shows the same trends with no changes in ozone in Atlanta since 1980.[47]

The number of "smog days" in Atlanta and its trend are frequently overstated. Seabrook (1999a) says about Atlanta, "But people with ailments are not the only ones whose lives and lifestyles

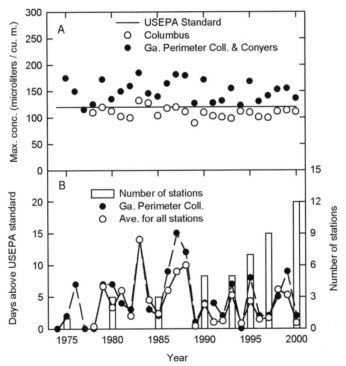

Figure 5.7. Maximum ozone concentrations for two stations in metropolitan Atlanta (averaged) and for Columbus from the 1970s to 2000 (A) and the number of days each year over the EPA standard of 120 microliters per cubic meter for Georgia Perimeter College and the metropolitan area (B). Bars show the number of ozone measuring stations in the Atlanta area. The horizontal line in (A) is the USEPA standard. From the US Environmental Protection Agency (1973–1998, 1994–2001), and the Georgia Department of Natural Resources (1985–2000).

are affected when smog levels rise to unhealthy levels, as they do on more than 40 days every summer." An editorial in the *Atlanta Constitution* on 16 September 1999[48] indicated that high ozone days had increased in the 1990s and gave the number of days as sixty for 1998 and sixty-seven for 1999. Perhaps the greatest overstatement was in the *Athens Daily News-Banner Herald*[49] which contained the following: "In that same period, the number of bad air days in Atlanta has more than quintupled from 13 days in 1994 to more than 60 days this year." In 1990 there were fifty-two days over the standard that Shearer used for comparison (new standard—see below).[50] So, it could be said that "smog days" decreased by 75 percent from 1990 to 1994.

The statements in the *Atlanta Journal-Constitution* and the *Athens Daily News-Banner Herald* are for days above a new standard of 80 microliters per cubic meter averaged over an eight-hour period that the USEPA proposed in 1997. The newspaper articles do not make clear, however, whether the old or new standard is being used. St. John and Chameides (1997) analyzed the days in Atlanta surpassing the two standards for 1987 through 1993 and concluded that the newly proposed standard would have been exceeded three times more often than the older one-hour standard. It is certainly not surprising that Atlanta air exceeds ozone standards more often when the standards are lowered. It is probable that other Georgia locations will also exceed the proposed lower standard. Macon, Columbus, and Augusta "are very concerned" about high ozone levels according to Dr. William Chameides.[51]

The counting of "ozone days" days is also subject to another kind of inflation, which comes from the addition of more monitoring stations in the Atlanta area in recent years. The USEPA ozone standard is violated if one station in the thirteen-county "ozone nonattainment area" measures levels above 120 microliters per cubic meter for one hour, or according to the new standard averages above 80 microliters per cubic meter for eight hours. The spotty nature of the ozone patterns in Atlanta means that some stations do not detect the highest concentrations in the metropolitan area. As the number of stations increase, so do the chances of detecting ozone concentrations above the standard. The number of stations in metropolitan Atlanta has increased from one in 1975 to twelve in 2000 (Figure 5.7B). An example of the inflation caused by more stations is that six days were above the 120-microliter-per-liter standard in 1996 for three Atlanta "trend sites," but including six stations in Atlanta doubled the number of days over the standard.[52] Even with the increased number of stations in the metropolitan area, the number of days above the standard was similar in the 1980s (11.8 days per year) and the 1990s (12.2 days per year). If inflation is avoided by considering only a single station or by averaging the number days above the standard for individual stations across the metropolitan area (Figure 5.7B), it is even clearer that the number of high ozone days has not increased.

Perhaps the perception that air quality in Atlanta is getting worse comes from the expectation that rapid growth in population and automobile numbers are bound to cause more polluted air. The population and the number of cars in the Atlanta ozone nonattainment area approximately doubled from 1975 to 1995. If one experiences mainly the interstate highways and other thoroughfares with their ever increasing traffic and ever longer waits at intersections, auto fumes and frustration must bias the perception of pollution. It is not likely, however, that the commuter carries the perception of pollution on Interstate 285 home to the subdivision in Roswell or Decatur. Of course the neighborhood air is not as polluted as the roadway lined with cars.

But roadways make up a small part of the metropolitan area, and the perception of increased population density may be more apparent than real. The population in the metropolitan area increased 41 percent, from 1.3 million in 1982 to 1.9 million in 1992 according to McCrary and Kundell (1997). However, they also estimated that the metropolitan land area increased by about the same ratio (44 percent) during the ten-year period. The result is that there was about the same density of people in 1982 (2.17 per acre) as in 1992 (2.13 per acre). The US Bureau of the Census (1920–2000) using an area of 6,126.2 square miles and a population of 3,627,184 for the metropolitan statistical area calculated a density of 592.1 people per square mile or slightly less than one per acre in 1997. These densities seem low unless it is realized that more than half of the metropolitan area is in forest.[53] Perhaps one reason for the lack of increase in ozone in the 1980s and 1990s is the fact that the Atlanta metropolitan area has gotten larger preventing even more crowding by people and cars. Fulton County increased from 1.78 people per acre in 1970 to 1.91 in 1990 and Dekalb County increased from 2.66 to 3.11 people per acre during the same period. The spreading of the metropolitan area or "urban sprawl," which is so widely cursed, may be preventing a worse air pollution problem that might result if growth were confined to two or three counties.

The main mitigation of air pollution in Atlanta and other urban areas has not been the control of population density, but the more efficient use of gasoline and cleaner exhaust of cars. The average efficiency of new cars and light trucks has almost doubled from 13 miles per gallon in 1970 to 25 miles per gallon in 1995.[54] There is also less production per car of the two main pollutants that result in ozone formation. Government mandated emission limits on nitrogen oxides from new car exhaust was reduced 87 percent from 1972 to 1994 (from 3.1 to 0.4 grams per mile).[55] A similar reduction was mandated for hydrocarbons, a component of the volatile organic compounds in the air. Increased fuel efficiency and reduction of emissions have helped overcome increased pollution that might have resulted from the rise in cars and miles driven in metropolitan Atlanta. The reduction in emissions is likely to be somewhat less than the government mandates, however, because of the number of older cars on the roads and the failure of emission controls with age. Regular inspections, which are required in some counties, should keep failures to a minimum.

The problem of what further to do about ozone is apparently complicated, not only by the habit of urban dwellers to drive one car per person, but also by the lack of knowledge of what controls the ozone level. While the basic chemical reactions that produce ozone are generally known, the practical cause and the solution of the ozone problem are not. The experts are fairly certain that the problem would be solved by the elimination of a large proportion of the cars and smokestacks in Atlanta, but have been less certain about which of the pollutants cause the problem. It appears that two main classes of pollutants are to blame, nitrogen oxides and volatile organic compounds.

In the 1970s and 1980s, the large amount of volatile organic compounds put into the air was considered the main cause of high ozone levels. The USEPA mandated a reduction in the release of VOC, and according to the Georgia Department of Natural Resources (1988) VOC releases in Atlanta dropped by 50 percent from 1977 to 1988. An estimated $750 million were spent during that period on VOC emission controls, mostly by industries. The lack of a reduction of ozone following the reduction of VOC prompted a reexamination of the problem. As concluded by Chameides and Cowling (1995), "After more than 15 years of progressively tighter and tighter controls on emissions of ozone precursors (especially VOC), little documented progress (in reducing ozone concentrations) [words in parenthesis are mine] has been achieved."

One hitch in solving the ozone problem by reducing VOC is that trees are a prime source of these substances. According to Dr. Chameides, as much as 65 percent of the VOC in the southeastern United States comes from trees.[56] Geron et al (1995), estimated in 1990 that 59 percent of the eleven-county Atlanta metropolitan area was covered by forest. Actually this is likely an underestimate of trees in Atlanta because not all the trees are in forests (see Figure 3.8). Even with 59 percent forest cover, they concluded that because so much man-made NOx was in Atlanta air, trees could contribute enough VOC to account for the ozone levels even if there were no man-made sources of VOC. The evidence that reducing man-made VOC had no effect on ozone and that trees produce a large proportion of the VOC, indicates that the other main contributor to ozone production, NOx, is controlling the level.

The role of NOx in controlling ozone was the subject of much debate in the 1990s. Nitrogen oxide emissions were estimated to have increased steadily during the first seventy-five years of the twentieth century, and were about three times as high in the 1970s as in the 1920s.[57] Although most of the NOx put into the air comes from combustion of fossil fuels, the nitrogen does not come directly from the fuel but from the air, combining with oxygen at high temperatures in engines and furnaces.[58] Because nitrogen oxides are formed in the air at high temperatures, lightning, forest fires, and other extremely high temperatures also produce them. This is a main reason for the ban on open burning of forest land and other debris in metro and surrounding counties during the summer months.

Although NOx is apparently high enough in Atlanta during hot, stagnant-air days in the summer to cause high ozone levels, it has actually decreased during the 1980s and 1990s. Nitrogen oxides have been measured only at Georgia Tech and in Dekalb County for more than a few years. If the measurements presented in Figure 5.8A are accurate, there has been a marked decrease in NOx in Atlanta since the 1970s. Measurements before the 1980s at the Fulton County Health Department may be suspect because of the poorer methods then, but even since 1983 there has been a real decrease of 12 percent in the yearly average NOx level at Georgia Tech. The USEPA also reported a downward trend for annual NOx levels in Atlanta for the period 1986–1997.[59] So, if high levels of NOx cause high ozone, the driving force for increased ozone during the 1980s and 1990s is lacking, because NOx has not increased. This lack of increase is remarkable because of the large increase in the number of cars that produce NOx.

As in the case of sulfur dioxide, the trends for NOx in the air and estimated emissions of NOx do not agree. Between 1980 and 1998 the estimated emissions of NOx in Georgia increased by 24 percent from about 590,000 to 730,000 tons, while measured concentrations either did not change or decreased (Figure 5.8). The lack of agreement of the trends may be because emissions are estimated for the whole state and the air levels are measured only in Atlanta. Or, perhaps it is because NOx was measured only at two locations in Atlanta and increasingly, NOx emissions from cars are spread

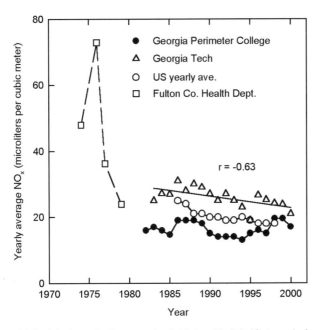

Figure 5.8. Trends in nitrogen dioxide concentrations in Atlanta and the United States. r-value for concentrations at Georgia Tech is the correlation coefficient showing a significant decrease. From the US Environmental Protection Agency (1973–1998, 1994–2001), and the Georgia Department of Natural Resources (1985–2000).

over a larger metro area. These are unlikely to be the reasons, however, because the same lack of agreement between emission and air concentration trends has occurred at the national level. The estimated national emissions of NOx were the same in 1988 and 1994, but NOx levels measured in the air at 205 stations across the United States decreased by about 17 percent from 1988 to 1995.[60]

Lead Poisoning

Unlike NOx, which comes from the air not gasoline, lead is a pollutant that was added to gasoline for over fifty years and was discharged in the exhaust with other pollutants. The lead-hydrocarbon compounds tetraethyl lead and tetramethyl lead were used as additives in gasoline to prevent knocking or "pinging" of engines that require high-octane fuel. Addition of lead was cheaper than boosting the octane rating by extensive refining. Use of these additives increased steadily until the mid-1970s when over 200,000 tons of lead were used in gasoline in the United States. In large cities, concentrations of lead in the air reached 2 to 4 micrograms per cubic meter, compared to values as low as 0.2 microgram per cubic meter in suburbs, and less in rural areas. These are lead levels at monitoring stations; concentrations in streets or highways with heavy traffic may reach 14 to 25 micrograms per cubic meter.[61]

Lead poisoning results from the breathing of lead in the air or otherwise consuming lead. The symptoms of lead poisoning are variously listed as anemia, irritability, weight loss, muscle cramps, stomach cramps and sluggishness. Although I have chosen to include lead as an air pollutant because so much of it was put into the air from car exhaust, it has been consumed more often throughout history in food and drink. The ancient Romans used lead pipes and solder to build water lines and this use continued past the mid-twentieth century, at least for soldering of water lines. If lead-containing solder is used in water lines, a small amount of lead is always in the water in those lines. The amount of lead dissolved from the solder depends on minerals or other chemicals in the water, some being more reactive with lead than others.

I remember rumors when I was young of people suffering lead poisoning from drinking moonshine whisky distilled using car radiators as the condenser for the alcohol. Lead in moonshine was never as serious a problem as the moonshine itself, but it is true that lead has been used to seal many metallic containers, from car radiators to soft drink cans. It is the softest and heaviest of the common metals and its low melting point led to its widespread use in sealing metal joints. Lead has also been used in many other products, such as pesticides, paint, stained-glass windows, batteries, and electrical cable sheaths, to name a few. Over 3 million tons of lead are mined in the world every year.

Because of the widespread concern about the health hazard of lead in the environment, the use of lead and its release to the air, land, and water has been reduced. In the 1970s, the use of lead additives in gasoline was restricted and has since been nearly eliminated. Lead additions to gasoline were eliminated for two reasons, to reduce the health hazard of lead in the environment and to prevent damage to catalytic converters installed in exhaust systems to reduce the emission of other pollutants.[62] The USEPA estimated that lead put into the nation's air by cars and trucks decreased from 181,698 tons in 1970 to 18 tons in 1997, a decrease of more than 99 percent. Lead loss to the air from metal processing and other industries was reduced 92 percent during the same period. Forty-seven percent of food and soft-drink cans were sealed with lead solder in 1981; by 1991 lead soldered cans were no longer made in the United States. About 83 percent of private homes and 86 percent of public housing units built before 1980 contain some lead-based paint;[63] but since then the use of lead in paint has been nearly eliminated.

As a result of the decreased use of lead in gasoline, lead in the air is only a small fraction of that in the 1970s (Figure 5.9). In the period, 1970 to 1979, lead levels averaged over the highest quarter of the year, varied from about 1.5 to 2.0 micrograms per cubic meter in Atlanta and about 0.75 to 1.0 microgram per cubic meter at three to eight other stations throughout Georgia. Beginning in the late 1970s, however, lead in the air began a steep decrease that reached nearly zero in the 1990s.

An exception to the downward trend of lead in Georgia air since the 1970s is the unusual pattern for two stations in Columbus. Columbus had one station measuring lead from 1970 to 1990 and the lead dropped from values near 1.0 microgram per cubic meter in the 1970s to very low levels in the 1990s. In 1991, two new stations were set up near a lead battery plant. At the startup of the new stations lead concentrations averaged about 2.0 micrograms per cubic meter for the highest quarter of the year. Because of these

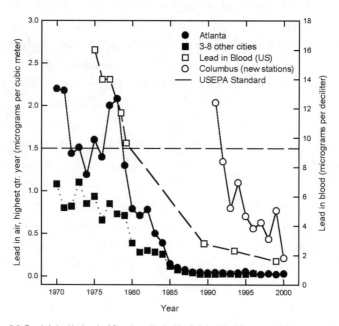

Figure 5.9. Trends in lead in the air of Georgia and in the blood of the United States population. Data for air from the US Environmental Protection Agency (1973–1998, 1994–2001), and the Georgia Department of Natural Resources (1985–2000); data for blood from Annest et al. (1982), Pirkle et al. (1994, 1998), and Centers for Disease Control (2001).

results Columbus was listed as a nonattainment area for lead. Nine years later the levels were decreased by 90 percent because of improvements in the plant and decreased production.[64]

Although separate figures are not available for Georgia, lead in the blood of tens of thousands of persons (9,632 from 1976 to 1980; 13,642 from 1991 to 1994) across the United States dropped 90 percent from the mid-1970s to the end of the century (Figure 5.9).[65] Such a reduction in blood lead levels is testament to the effectiveness of public opinion and technology in solving pollution problems in the latter half of the twentieth century. Although James O. Mason of the US Public Health Service said in 1991, "Lead poisoning is one of the most common pediatric health problems in the United States today, and it is entirely preventable,"[66] it is certainly nowhere near the problem it was in the 1970s. Actually, blood lead levels in children one to five years old dropped by about the same amount as shown for the general population in Figure 5.9.[67] It is likely that decreased lead in the blood is mainly due to removing lead from the air, but removal from food and drink cans, paint, water lines, and other products certainly helped.

Carbon Monoxide

Carbon monoxide is a poisonous gas, as most people know from having heard of successful or attempted suicides using exhaust from automobiles. There are also occasional reports of sickness or death caused by carbon monoxide from the malfunction of home heaters. When fuel burns completely, the carbon is oxidized to carbon dioxide (CO_2). When combustion is incomplete, some of the carbon is incompletely oxidized and released as carbon monoxide (CO). Because combustion of fuel is seldom 100 percent complete, there is almost always some carbon monoxide associated with flames. The concentration is usually low enough to be safe, but under some conditions of incomplete combustion and poor ventilation it increases to dangerous levels.

When inhaled, carbon monoxide combines with hemoglobin in the blood to form carboxyhemoglobin, reducing its ability to carry oxygen. The shortage of oxygen supply to the body causes sickness and death. By far the most common cause of carboxyhemoglobin formation in people is the smoking of tobacco and inhaling tobacco smoke. Typically cigarette smokers have a carboxyhemoglobin level of 5 percent of their total hemoglobin compared to about 1 percent in non-smokers. However, traffic policemen, garage attendants, and drivers of taxis and buses experience increases to about 3 percent. Levels above 10 percent can cause fatigue, drowsiness, reduced work capacity, coma, respiratory failure, and eventually death.[68]

Although carbon monoxide can build up to dangerous levels in confined spaces such as garages and homes, it seldom accumulates to dangerous levels in the open. Carbon monoxide levels in urban areas show daily peaks at the morning and afternoon traffic rush hours. Concentrations are generally lower than 17 milliliters (1,000 microliters) per cubic meter, but may be higher in tunnels, underpasses, and other confined areas.[69] Because of the danger of carbon monoxide and the large production of it by gasoline engines and factory furnaces, this gas was included in the USEPA list of "criteria gases" (Table 5.1) and subjected to regulation. Note, however, that the USEPA standard concentration is 50 to more than 100 times higher than for the other gaseous pollutants.

Monitoring in Georgia began in the 1970s with five stations in Fulton County. However, the perceived danger must have been much less than for other pollutants because since 1985 measurements have continued at only two stations, both in Fulton County. The average of the second-highest eight-hour levels of carbon monoxide each year in the Atlanta area have decreased from about 9 milliliters per cubic meter in 1978 to 2.8 milliliters per cubic meter in 2000, a drop of 69 percent (Figure 5.10). The second-highest eight-hour levels are presented for comparison to the US

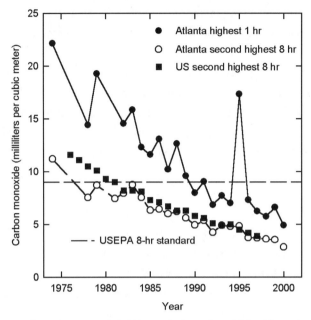

Figure 5.10. Trends since the 1970s in the highest yearly 1-hour and second highest 8-hour carbon monoxide concentrations in Atlanta and the second highest 8-hour concentrations for the United States. From the US Environmental Protection Agency (1973–1998, 1994–2001), and the Georgia Department of Natural Resources (1985–2000).

trends, which are the measurements given in the National Air Quality and Emissions Trends Reports. Levels in Atlanta have followed the national average closely. The highest one-hour averages are also given for Fulton County and they follow a similar trend, but not so smoothly.

Carbon monoxide in Fulton County never exceeded the 35-milliliter-per-cubic-meter 1-hour USEPA standard given in Table 5.1 and the second highest eight-hour average did not exceed the 9 milliliter-per-cubic-meter standard after 1980 (Figure 5.10). In 1978 and 1979, seven of the ten highest individual values reported from the five stations in Fulton County were over the eight-hour standard. Although a few individual eight-hour values exceeded the standard in the early 1980s, none did in the 1990s. The decrease of carbon monoxide in the air since the 1970s is apparently due to pollution controls on cars. Catalytic converters in the exhaust systems scrub out most of the carbon monoxide that used to escape into the air.

The decreasing trend of carbon monoxide concentrations in Atlanta does not match the changes of emissions estimated by the USEPA for Georgia. According to these estimates, carbon monoxide release dropped from 2.5 million tons in 1970 to 1.7 million tons in 1986 and then increased again to 3.9 million tons in 1997. If these figures are accurate, there should have been an increase in concentrations in the 1980s and 1990s. It is possible that carbon monoxide releases in other locations increased and that in Atlanta continued to decrease. Such a change in patterns of emissions is unlikely however. Because emissions are calculated based on kinds and amounts of fuels used, and use and efficiency of pollution control devices on automobiles and factory and power station stacks, there is room for error in emissions calculations.

Miscellaneous Pollutants

There are many air pollutants not discussed in this chapter because they are rare in occurrence, minor in their effects, or there are too few data to make any conclusions about their trends or danger. There are a number of organic vapors that are released from industrial plants and are regulated by the USEPA, some as toxic chemicals. Many of the gases released to the air in the toxic category are not very toxic. For example 40 percent of the total weight of toxic materials released to Georgia air from stationary sources in 1996 was methanol and ammonia mostly from wood processing plants.[70] Three of the other top ten chemicals released to the air were vapors of the solvents, styrene, toluene, and xylene, certainly not healthy vapors if concentrations are high, but also not known for health problems in Georgia. Problems with these chemicals are not widespread enough for them to be routinely measured in air by the Georgia Environmental Protection Division. Regardless of the toxicity of these gases and the relatively small problems they cause, their releases to the air in Georgia have decreased from 95 million pounds in 1985 to 35 million in 1996.[71]

There is considerable concern about gases that accumulate in the atmosphere and absorb infrared rays from the earth contributing to what is called "the greenhouse effect." The gas of main concern is carbon dioxide. It is questionable if carbon dioxide is a pollutant. Wilson (1996) defined a pollutant as "any material entering the environment that has undesired effects." The definition seems to be incomplete if the material also has beneficial effects. If a material has beneficial effects, its pollution potential depends on detrimental effects outweighing beneficial effects. At very high concentrations, carbon dioxide is toxic, but at levels now in the air and those predicted for the future, it is not close to being dangerous for human health. If you inhale while drinking a freshly poured carbonated drink, the level of carbon dioxide inhaled is probably 100 times the normal air concentration, and certainly no adverse effects are felt.

Carbon dioxide enters the air mainly from two sources, respiration by plants and animals and the burning of fossil fuels and wood. That placed in the air by burning of fuel and forests is not dissipated quickly like many pollutants, but remains in the air until assimilated by photosynthesis in green plants. Presumably before the industrial revolution, there was a rough balance between the amount of carbon dioxide released by plants and animals and that fixed by plants. Since the beginning of the industrial age, carbon dioxide has been released at a faster rate than plants assimilate it, and it is accumulating in the air. It is estimated that the pre-industrial (early 1800s) concentrations of carbon dioxide were about 280 microliters per liter. At the end of the twentieth century, the concentrations are about 350–360 parts per million, a 25–30 percent increase.

So the main concern for rising carbon dioxide levels is the warming of the earth, the "greenhouse effect." The detrimental effects of carbon dioxide as a "greenhouse gas," however, are speculative, controversial, and in the future. There is no question that carbon dioxide is a greenhouse gas, but in fact greenhouse gases are absolutely essential for survival of the earth, as we know it. These gases moderate the temperature of the earth to keep it from fluctuating hundreds of degrees as it does on planets with little or no atmospheric gases that absorb infrared radiation. In addition, nearly all life on earth depends on photosynthesis of green plants and carbon dioxide is the main raw material assimilated in the process. The concern about carbon dioxide is that increases above historic levels will cause an increase of a few degrees in the earth's temperature and severe disruptions of weather patterns.

Some scientists think that the rising carbon dioxide levels have already had an effect on the earth. Reports in the 1990s indicate that the temperature of the earth has already increased a fraction of a degree in the last 100 years. The main argument seems to be whether the increase is real, and if it is, whether rising carbon dioxide caused it. The concern about global temperature and other weather changes is strong enough that it has spawned many national and international conferences over the last thirty years to consider the dangers and solutions to the rise in carbon dioxide.

Accumulation of carbon dioxide in the air is not all bad because higher concentrations cause plants to grow faster. In the range of 280–350 microliters per liter of carbon dioxide, the concentrations existing over the last 200 years, an increase in carbon dioxide causes a nearly proportional increase in photosynthesis. That is, there is a near doubling of photosynthesis for a doubling of carbon dioxide. In some plants, photosynthesis increases with carbon dioxide up to two or three times the present levels, but in others photosynthesis is nearly maximum at current levels. It has been demonstrated that many crops will yield more and forest trees will grow faster at carbon dioxide levels predicted for the future. So the increases in carbon dioxide of the past and future will certainly have some beneficial effects and possibly negative ones, and time will tell what the balance will be.

Visibility

One manifestation of clean air is long distance visibility; the clearer the air, the further one can see. The haze over cities and even the haze that makes the Blue Ridge Mountains blue are caused by particles and some gases in the air. Although visibility is reduced in both cases, we acclaim the haze of the Blue Ridge because it gives a special beauty to the ridges and because it is largely natural, but detest the city haze because it is pollution that burns our eyes and makes us cough. If air pollution has changed much during the last half-century, it would be expected to affect visibility.

In fact, some studies have indicated that visibility in the southeastern United States has decreased in the last half of the twentieth century. Trijonis (1986), Husar et al (1981), and Husar and Wilson (1993) concluded that average visibility in the southeastern United States has dropped by 25–50 percent and summertime haziness (calculated from visibility measurements) has increased by 80 percent. Husar and Wilson (1993) and Sloane (1984) attributed decreases in spring and summertime visibility in the mid-eastern United States from the 1950s through the 1970s to increases in sulfur dioxide emissions caused by increased coal usage during the period. The decreased visibility has been attributed to increases in small sulfur-containing particles in the air. Particles with diameters smaller than 2.5 micrometers in the southeastern United States are high in sulfates, which usually makes up one-half or more of their weight.[72] Reports of decreased visibility for the southeast led to Kundell and Swanson's (1991) conclusion that visibility had decreased in Georgia and that higher sulfur dioxide emissions were the cause.

Although it is not the reduced visibility that is most annoying about city air, it is important at airports for landing and takeoff. And because it is, measurements have been made at airports for many years. In Georgia, the most complete records are available from Atlanta's Hartsfield International Airport. I obtained records back to 1948 for that airport and back to 1962 for Savannah, Augusta, and Macon. These records were examined for trends that may be related to changes in air pollution. The National Weather Service made the observations and those for Atlanta from 1948 to 1978 were taken from a US Department of Transportation (1981) summary. Records for all four cities from 1962 to 1995 were obtained in a summary form from Dr. R. B. Husar of Washington University in St. Louis. The maximum distance recorded on clear days depends not only on particles and vapor in the air, but also the distance of the farthest target available to the observer. Because the distance of the targets may have changed over time and because pollution is expected to reduce visibility, days of highest visibility have not been used to show trends with time. Instead, the percentages of observations with visibility below 7 miles are shown. The cleaner the air, the fewer days of visibility below 7 miles there should be.

Visibility at Atlanta airport has gone through a down and up cycle since the 1950s. The decreases in visibility reported for the southeast in the 1960s and 1970s are borne out in Figure 5.11A by the increased percentage of low-visibility observations from the 1950s to about 1980. However, since then the percentage appears to have decreased from an average of about fifteen in the 1970s and 1980s to eleven in the 1990s. The two sets of observations in Figure 5.11A obtained from the US Department of Transportation (1981) report and from Dr. R. B. Husar overlap from 1962 to 1978 and show the same upward trend for low-visibility days during that time. The similar upward trend is expected because the same set of Weather Bureau records is used, but the Department of Transportation numbers (triangles) are the percentages of all observations and those beginning in 1962 are percentages only of observations at noon and omit the days with rain and fog. The worsening of air quality in the southeast has been attributed mainly to reduced summertime visibility.[73] If the observations beginning in 1962 in Figure 5.11A are separated into June–August and September–May periods, there is little change in the summer visibility since the 1970s, but a much greater decrease in low-visibility days for September–May.

The increase in observations below 7 miles at the Atlanta Airport from mid-century to the 1970s may be related to the increased air pollution in the area during that time because of the growth of that area. When a new terminal was completed in 1961,

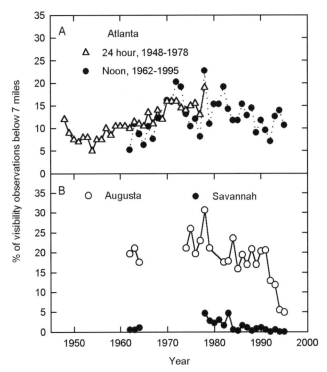

Figure 5.11. Trends in the percentage of visibility observations below 7 miles for airports at Atlanta (A) and Savannah and Augusta (B). Gaps are for years of no records. Records for 1948 to 1978 in Atlanta (triangles) are from the US Department of Transportation (1981) and include all observations. Other records supplied by Dr. R. B. Husar, Washington University are for observations made only at noon and periods of fog and rain were excluded.

the largest single terminal building in the country, it encouraged or facilitated a rapid growth of the area that was then relatively undeveloped. Clayton County, where the airport is located, and the other three surrounding counties, Fulton, Dekalb, and Henry had a combined population of 649,000 people in 1950, but 1.25 million in 1980, an increase of 94 percent. Car registrations for the four counties increased 71 percent, from 516,000 in 1965 to 883,000 in 1980.[74] The increase in low visibility percentages from the 1950s to the 1970s also follows fairly closely the emissions of sulfur dioxide calculated for Georgia in the same period.[75] In addition, emissions peaked about 1985 and decreased in rough agreement with the improvement of visibility at the Atlanta airport.

Although low-visibility days reached a maximum and declined after the 1970s (Figure 5.11A), expansion of the airport continued. The number of passengers continued to increase in a near linear fashion from about 3.8 million in 1961[76] to over 31 million in 1997.[77] The number of flights and the size of airplanes increased in parallel with the number of passengers. Jet planes were few and small in the 1960s, compared to those that carry the tens of millions of passengers today. The population of the four-county area around the airport has increased another 30 percent since 1980. So growth of the airport and the spreading of the Atlanta metropolitan area to engulf the airport brought the people, industry, and traffic that would be expected to worsen its air pollution. However as noted earlier, sulfur dioxide, the main form of sulfur emissions have declined in Atlanta air since the first measurements in the mid-1970s (Figure 5.2) and particles in the air have decreased since the mid-1950s (Figure 5.1). So the air quality in the Atlanta area began and continued to improve for at least twenty years before the low-visibility observations started to decline.

The records for Augusta and Savannah are sparse for the early years in Figure 5.11B, but they show the same downward trend of low-visibility observations as Atlanta during the 1980s and 1990s.

Although the percentage of low-visibility observations for Augusta was similar to or slightly higher than for Atlanta in the 1970s and 1980s, they dropped sharply to around 5 percent in the 1990s. Days of visibility below 7 miles at noon were never very high in Savannah. They peaked at about 5 percent (about eighteen days per year) in the late 1970s and early 1980s and decreased to nearly zero in the 1980s and 1990s. Visibility for Macon is not shown in Figure 5.11, but the values and trends were very similar to those for Augusta. Taking the observations for the four airports together it appears that although visibility in Georgia may have grown worse from some earlier time to the 1970s, it has improved since then.

Air Pollution and Health

It is not in the purpose of this book to examine how the different pollutants cause illness and at what levels. However, it is worthwhile to ask whether the clearing of the air has improved health. The specific answer to the question is mostly lost in uncertainty—uncertainty about which air pollutants are responsible for illness, about whether some unhealthy components are not measured, and uncertainty about the validity of associations between pollution and illness. Nearly all studies of air pollution and illness are associative. That is, rates of sickness or death are compared to differences in concentration of one or more pollutants measured in different cities, neighborhoods, or time periods. If rates of death or sickness are higher in polluted neighborhoods or in periods of high pollution, it is concluded that pollution contributed to the increased illness.

The answer to the question about whether improvements in air quality have improved health is also obscured by changes in the prevalence of respiratory disease and cures that are unrelated to air pollution. New drugs, inhalants, and better diagnosis have improved respiratory treatment over the same period of time that the air has become cleaner. So any decreases in sickness or death due to respiratory diseases may be due to improvements in treatment rather than or in addition to cleaner air.

Country air has always been seen as cleaner than city air and the air quality measurements presented in this chapter bear that out for Georgia. City residents have always been considered more prone to air-quality related ailments than rural people. Over time, however, the differences have diminished. Air has become cleaner in both city and country, but the change in city air has been greater because it was worse at the start of the clean up. In the United States from 1950 to 1954, lung cancer was twice as common in cities as in rural areas, but by 1970 to 1975 it was only 13 percent higher.[78] An indication of the extent of cleaning of the air is that the second leading cause of lung cancer behind tobacco smoke is no longer outdoor pollutants, but radon in homes.[79]

On the other hand, there are claims that asthma and other respiratory problems are increasing in Georgia.[80] Although the commonly measured pollutants have decreased markedly or, in the case of ozone, remained constant in Georgia air, there is speculation about the increasing detrimental effects of "fine particulates" in the air.[81] In the book, "Particles in Our Air,"[82] it is concluded that health risk increases as inhaled particles become smaller, because smaller particles penetrate deeper into the lungs and contain more reactive chemicals. The frequent references to decreasing visibility in the Southeast and its linkage to "fine particle" sulfate suggests an increase in fine particles high in sulfur. The region of the Appalachian Mountains has the largest proportion of sulfates in "fine particulates" of the air.[83] But sulfate particles of any kind are unlikely to have increased because of the large decrease in sulfur dioxide in the air (Figure 5.2, Table 5.3). If asthma or other respiratory diseases have increased in Georgia in the last quarter-century, the cause is unlikely to be air pollution, which has decreased.

Asthma is not even concentrated in Georgia counties with highest air pollution. A recent study concluded about asthma-related hospitalizations that "Although counties with high rates are located in all parts of Georgia, they are more common in a band extending from Augusta to the southwest corner of the state."[84] Atlanta is outside this band.

One of the problems in associating air pollutants with respiratory illness is determining the level of exposure of human subjects. The levels of pollutants at monitoring stations represent poorly the exposure of humans, because pollutants are not measured where people live, especially indoors where most people spend most of their time. Some buildings have pollutant levels below that of outdoors, but others may have higher levels than outdoors. Cooking, smoking, fireplaces, and wood heaters may cause pollutants to be higher indoors than out. Likewise, pet dander, cockroaches, and other house dust sources may also create more pollution than outside. So, driving to the office, working eight hours, driving home, smoking a pack of cigarettes, sweeping the garage, mowing the lawn, cooking dinner, and sitting by the fireplace exposes a person to an unknown dosage of pollutants that is certainly not represented by monitoring stations. This poor representation of human dosage at monitoring stations probably explains why most studies show weak associations between pollutant concentrations and respiratory ailments.[85]

In spite of the fact that monitoring stations do not represent pollutant doses well and in spite of the low associations between pollutant concentrations and sickness, there have now been hundreds of studies that show air pollution is related to respiratory illness. There are in fact enough studies to make a convincing case that heavy air pollution is associated with increased illness. But if the hundreds of studies prove that association, then reduction of four of the six criteria pollutants by more than 50 percent and no increase in the others proves that Georgia's air is now the healthiest in at least forty years. The data that prove the point are not hand-picked, and they are not restricted to Georgia. The 1996 Air

Table 5.6. Long-term Changes in National Air Quality Concentrations and Emissions. From the 1996 Air Quality and Emissions Trends Report (U.S. Environmental Protection Agency, 1973-1998).

	Air Quality Concentration % Change[1] 1977-1996	Emissions % Change[1] 1970-1996
Carbon Monoxide	−61%	−31%
Lead	−97%	−98%
Nitrogen Dioxide	−27%	+8%
Ozone	−30%	−38%
PM-10	−[3]	−73%[4]
Sulfur Dioxide	−58%	−39%

[1] A minus sign shows decreased pollutants. [2] Emissions are calculated for all nitrogen oxides (NO$_X$) and for ozone, all volatile organic compounds (VOC). [3] PM-10 was not measured before the mid-1980s. [4] Includes only directly emitted particles. Secondary particles formed from pollutant gases make up a significant fraction of PM-10.

Quality and Emissions Trends Report from the USEPA contained a table that is reproduced as Table 5.6. It shows that for the United States as a whole, every criteria pollutant has decreased since 1977, with decreases ranging from 27 to 97 percent. The great good news about cleaner air is undeniable.

[1] Kuhn et al, 1990.
[2] Wilson, 1996.
[3] Wilson, 1996.
[4] Schrenk et al, 1949.
[5] Weyandt, 1998.
[6] Range, 1954.
[7] Husar, 1986.
[8] Citizens Fact-Finding Movement of Georgia, 1946.
[9] Kuhn et al, 1990.
[10] Kuhn et al, 1990.
[11] Reynolds and Pierson, 1942.
[12] Citizens Fact-Finding Movement of Georgia, 1946.
[13] US Environmental Protection Agency, 1996.
[14] Citizens Fact-Finding Movement of Georgia, 1946.
[15] Spillers and Eldredge, 1943.
[16] Georgia Department of Community Affairs, 1997.
[17] Carter, 2001.
[18] US Environmental Protection Agency, 1994.
[19] US Bureau of the Census, 1920–2000.
[20] Georgia Conservancy, 1979.
[21] Lynn et al, 1976.
[22] Spengler and Wilson, 1996.
[23] US Environmental Protection Agency, 1973–1998, 1994–2001.
[24] Seabrook, 2000c.
[25] Kundell, 1984.
[26] Kundell, 1984.
[27] US Environmental Protection Agency, 1973-1977.
[28] Gschwandtner et al, 1985.
[29] US Environmental Protection Agency, 1973-1998.
[30] Kundell, 1984.
[31] National Clean Air Coalition and Friends of the Earth Foundation, 1984.
[32] Rykiel, 1977.
[33] Kundell, 1984.
[34] National Clean Air Coalition and Friends of the Earth Foundation, 1984.
[35] Kundell, 1984.
[36] Landau, 1967.
[37] Harmon, 1990.
[38] Walker, 1982.
[39] Huntington, 1996; Lindberg and Lovett, 1992.
[40] Cappellato et al, 1998.
[41] Smith and Alexander, 1983.
[42] Fowler, 1985.
[43] Chameides et al, 1988.
[44] St. John and Chameides, 1997.
[45] St. John and Chameides, 1997.
[46] US Environmental Protection Agency, 1973-1998, 1994-2001.
[47] Georgia Department of Natural Resources, 1999b.
[48] Atlanta Constitution, 1999.
[49] Shearer, 1999.
[50] US Environmental Protection Agency, 1973–1998.
[51] Shearer, 1999.
[52] US Environmental Protection Agency, 1973-1998.

[53] Geron et al, 1995.
[54] Davis, 1997.
[55] Davis, 1997.
[56] Shearer, 1999.
[57] Gschwandtner et al, 1985.
[58] Husar, 1986.
[59] US Environmental Protection Agency, 1973–1998.
[60] Darlington et al, 1997.
[61] Elsom, 1987.
[62] Elsom, 1987.
[63] Pirkle et al, 1998.
[64] R. Ballagas, Georgia Department of Natural Resources, personal communication.
[65] See also Annest et al, 1982, Pirkle et al, 1994, 1998, Centers for Disease Control, 2001.
[66] Centers for Disease Control, 1991.
[67] Pirkle et al, 1994, 1998.
[68] Elsom, 1987.
[69] Elsom, 1987.
[70] US Environmental Protection Agency, 1999c.
[71] Georgia Statistical Abstract, various years.
[72] Spengler and Wilson, 1996.
[73] Husar and Wilson, 1993.
[74] Weyandt, 1998.
[75] Gschwandtner et al, 1985.
[76] Braden and Hagan, 1989.
[77] Federal Aviation Administration, 1998.
[78] Greenberg, 1987.
[79] US Environmental Protection Agency, 1992.
[80] Seabrook, 1999a, 1999b.
[81] Seabrook, 1999a, 1999b.
[82] Koutrakis and Sioutas, 1996.
[83] Spengler and Wilson, 1996.
[84] Mellinger-Birdsong et al, 2000.
[85] Gamble, 1998.

CHAPTER SIX

Restocking the Wildlife

"The bluebird...is American idealism personified—a flying piece of the sky, a living poem, a crystal note, an emblem of nature's moral conscience"[1]

Wildlife in the Past

I know from experience that wildlife changed a great deal in the twentieth century. I grew up on a farm in middle Georgia from the mid-1930s to the 1950s. I explored quite a lot of the countryside near our farm with my brothers, and we would often have a dog and a gun along, hunting for rabbits, squirrels, quail, or whatever other game crossed our path. Many of the animals that today are taken for granted, and numerous enough to be a nuisance to farmers and sometimes to suburban neighborhoods, did not exist in our woods. I was an adult and away from home before I saw a deer in the wild and its no wonder: only twenty deer were estimated for the whole county (Laurens) when I was sixteen-years-old,[2] and they were probably in the Oconee River swamp. I also never saw a turkey or a beaver in the wild. We very seldom saw wild geese fly over, and they never stopped on our pond, although ducks occasionally did. Bobcat and coyote were animals that we knew only from books, movies, or tall tales by neighbors or kin.

There was game to hunt when I was growing up, plenty of rabbits, squirrels, fox, quail, and doves. We hunted opossum and raccoon with hounds that were family pets as well. The abundance of small game was due to the habitat provided by small farms. Briar patches and brush along fencerows, small wood lots, fields of broomsedge that were recently out of cultivation, and grain fields with weed seeds and scattered grain were good habitat for such game. In fact, some of the small game species have diminished over the last fifty to sixty years because of the decrease in cultivated and idle land. Dense forests are not good for quail, which need open, weedy areas for food and nesting.

The fish we caught in Buckhorn and Rocky Creeks were not the bass weighing several pounds that fishermen expect today. They were pickerel about six or eight inches long, bream smaller than your hand, or catfish usually a half pound or less. Sometimes we used snatch-hooks to catch suckers that were bigger but not as desirable. We didn't fish for bass until my father built a farm pond about 1950 and stocked it with both bream and largemouth bass. I remember fishermen talking about going to Big Sandy Creek and Chappell's Mill pond between Dublin and Irwinton and to the bend in the Ocmulgee River near Lumber City to fish for the big ones. Although some of the large lakes had been built by the 1950s, they were too far away for us to use.

It was not always that way. Wildlife was abundant when European settlers came and undoubtedly made up a large portion of the early settler's diet. In some parts of the state, it was apparently quite a long time after settlement before extensive, additional clearing of the forest. Much of the sustenance of the early settlers came from wild game or the livestock that subsisted on the grasses, nuts, and other feed furnished by the open, burned-over woodlands.

Although the early settlers probably used many of the animals, birds, and fish of the state, they likely concentrated on the larger ones that supplied meat for longer periods. By most accounts, deer and wild turkey were plentiful and herds of buffalo existed in some parts of the colony.[3] The wiregrass region of Georgia from its settlement until the mid-1800s furnished extensive grazing for sheep and cattle before the forest was cleared.[4] The livestock of the region had free-range (no fences) and probably co-existed with deer and other forest animals.

Many factors have affected the abundance and balance of Georgia's wildlife, but changes in the landscape probably had the greatest influence. The conversion of the Georgia landscape from forest to a patchwork of fields, idle land, and forest reduced the numbers of large animals and birds, but probably increased the smaller ones. Forest fragmentation endangers some forest dwelling wildlife, but favors many species, especially songbirds.[5] The briar patch that was the beloved home of Brer Rabbit in Joel Chandler Harris' Uncle Remus tales was a common part of the cotton patch agriculture and diversified farms of the 1800s and early 1900s. These briar patches were an important part of the habitat of the rabbit, but also quail, sparrows, the brown thrasher, and many other birds. Underneath the briar cover were field mice, rats, snakes and various other animals and insects that made up a complex ecosystem. In the middle of the twentieth century fencerows were a part of the habitat of this menagerie of birds, reptiles, and mammals.

Clean fencerows were the pride of the fastidious farmer, but weedy, bushy ones were the norm and added to the diversity of the small animals and birds. As Chapman et al (1950) observed, "Clean fence lines are of no value to farm game." Even the progression of the older rail fences with their zigzag pattern to the more modern wire fences was lamented as detrimental to small game.[6] Because of the zigzag line of the rail fence, a much wider strip was left uncultivated. Of the modern movement to clean wire fences, Johnson stated, "Such modernization, taken as a whole, has been as great a factor in depleting upland game as has the gun." Besides serving as breeding and feeding ground to many species, fencerows served as passageways for birds and small animals to move safely to sources of food, water, and neighboring populations.

Several large species almost disappeared from Georgia in the period of cotton agriculture because of land use changes and also because of unregulated hunting.[7] Those same species are now relatively plentiful. The resurgence of wildlife, especially forest game, is largely the result of four factors. (1) The population on farms, many of whom were hunters, decreased drastically by the 1970s as people moved to towns and cities.[8] (2) Much land formerly in row crops was diverted to forest either through planting of trees or natural reforestation. (3) There were concerted efforts by wildlife specialists to restock several of the game animals and birds that had completely or nearly disappeared early in the twentieth century. (4) Hunting became much more regulated than before. A significant factor in the resurgence of the larger game is that fewer were harvested late in the century than earlier when people hunted year round, at least when they had the need and inclination. Hunting of game is now highly regulated and for most game species, restricted to a few days or weeks of each year. However, the most important reason for resurgence of wildlife is likely the change in habitat caused by changed land use.

Improvement in wildlife resources in the state is somewhat difficult to document, partially because no records were kept of wildlife populations early in the twentieth century, and for most wildlife, there are still no records. The problem of wild game scarcity was widely recognized just before mid-century when organizations such as the Georgia Department of Natural Resources began to survey game species, restock certain animals and birds, and regulate activities that affect game populations. At mid-twentieth century Jenkins (1953) concluded that, "Georgia could become one of the very best states in the United States in regard to wildlife production because of the low density of human population, tremendous variation in habitat...huge timberland tracts, deep swamps, and residual game populations." The term "residual game populations" was this wildlife management veteran's assessment of the status of Georgia's wild game.

Documentation of improvement is also made difficult by the uncertainty about what constitutes "improvement." Whitney (1994) quotes Aldo Leopold as saying that "Wildlife is never destroyed...it is simply converted from one form to another." I take this to mean that when food and habitat are available, it will be used. If habitat and food supply change, wildlife doesn't disappear, but the mixture of species shifts toward those that can use it or compete best for it. The switching of one species of wild animals for another may be viewed as improvement by some groups of hunters and wildlife enthusiasts but not by others. Abundant deer is an improvement for deer hunters, but may be a loss for quail hunters if deer habitat has replaced weedy fields and fencerows. The increase in large game populations in Georgia has coincided with decreased hunting of small game. According to the Georgia Department of Natural Resources (1993), "Declines in hunter numbers, hunting pressure, and harvest for most small game species continue to decline, particularly quail and dove, to near alarming levels." The following section gives accounts for the more popular and best-counted species.

Game Species
Deer

Wildlife, including white-tailed deer, was plentiful when the white man came. Although there was little concern early in the colony for what seemed to be an inexhaustible resource, there were early regulations imposed on local hunting. In 1773, the Colonial Governing Board of Georgia ruled that anyone found hunting deer with a light was to receive thirty lashes "well laid" on the back.[9] Deer must have been a staple food for Indians and settlers alike. The clearing of Georgia for cultivation of crops reduced the extent of habitat and "during the years between 1790 and 1835 their range gradually decreased and the numbers of herds decreased accordingly." Apparently late in the nineteenth century there was concern about disappearance of game, because in 1893 a law was passed prohibiting the selling of game. District Forest Ranger Arthur Woody, who aided in setting up the Chattahoochee National Forest in the 1920s, stated that his father killed the last deer in the Northeast Georgia mountains in about 1895.[10] Jenkins states that the large majority of the native deer in this section of the state were taken with large packs of well-trained dogs.

During the period of maximum farm numbers, the deer population was almost wiped out in other sections of the state. In a 1937 survey, there were only 12,500 deer estimated in Georgia.[11] Seventy-two percent of those were in four counties along the coast (Bryan, Liberty, Long, and McIntosh) which never had substantial clearing and cultivation. Two-thirds of the counties (108) were listed as having none at all. Deer were restricted to protected pockets mostly on large reservations owned by corporations or wealthy individuals. Henry Ford owned a large plantation on the Ogeechee River at Richmond Hill where deer were protected from poaching. The large reservations supplied protection, but most were not managed to encourage proliferation of deer.

In mid-century efforts to restock Georgia's deer herds intensified, after a small beginning in the 1920s. Jenkins relates, "In 1928, Ranger Woody purchased from his own funds six deer and released them in what later became the Blue Ridge Game Management Area." In the following two years, the Forest Service stocked three other areas with fifty deer. All of the deer were brought from out of state. The first hunt in the Blue Ridge Game Management Area was held in 1940, twelve years after restocking began, and twenty-two bucks were taken. In 1950, 182 deer were taken in the original restocking area. There was an increase in Georgia's deer from 12,500 in 1937 to 33,370 in 1951, but two-thirds of the deer were still in the coastal strip of counties[12] and sixty-five counties, mostly in the Piedmont, had less than ten deer each. But even in the optimism of the 1950s, the hope for recovery was modest, especially in the Piedmont. Jenkins concluded that "Eventually the piedmont may support a good deer population as it did in the early days of settlement, but for the present much of it should be considered as essentially farm game habitat from the wildlife standpoint." "Farm game" was that considered more appropriate for the small farms, such as rabbits, squirrels, and quail. Chapman et al (1950) too, reckoned that "These larger animals (deer and turkey) cannot, as a rule, be raised and fostered through farm conservation programs. They require too great an expanse of protected land." It is interesting that both Allen (1948) and Jenkins (1953) thought that the mountains of North Georgia were too densely covered with mature timber in mid-century, with not enough open spaces, to support large populations of deer. However, Jenkins thought that Georgia could support "nearly 400,000 deer without conflicting with agriculture."

Surely no one then envisioned more than a million deer in Georgia. The increase in deer population is one of the great success stories of environmental change in Georgia in the twentieth century. Through the mid-1980s, the yearly increases were ever larger, with deer populations doubling about every seven years from the 1950s to the late 1980s (Figure 6.1A). Hunters killed ten times more deer in 1990 than existed in Georgia in mid-century (Figure 6.1). In that year, there were over 350,000 deer hunters in the state, a population as large as the fifty smallest Georgia counties combined. The number of hunters increased about 40 percent from the 1970s to the 1990s (Table 6.1) and the number of deer taken increased from about one for every four hunters in 1974 to three for every four hunters in 1990. In the mid-1970s, the numbers of deer and hunters were about equal, but by 1987 the hunters were outnumbered three to one, which probably accounts for the increase in deer taken per hunter (Figure 6.1B). Although the number of hunters appear to have leveled off in the 1990s at about 350,000, the number of deer killed increased from about 300,000 to about 450,000 per year during the decade. The high deer population and popularity of deer hunting means it dwarfs the other game in the state. In 1996, 80 percent of the hunters in the state were deer

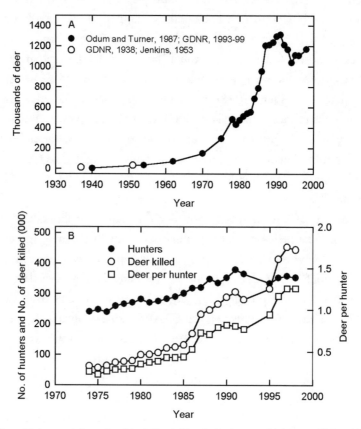

Figure 6.1. Increases in the number of deer in Georgia (A) and in deer hunters and their success (B). From Odum and Turner (1987), Jenkins (1953), and the Georgia Department of Natural Resources (1938, 1993–1999).

Table 6.1. Numbers of hunters (thousands) for various game in Georgia in the 1970s compared to the 1990s. From Georgia Department of Natural Resources, 1993, 1993-1999.

Game Species	1970s Years	1970s No. of hunters	1990s Years	1990s No. of hunters	% change
Wild turkey	74-79 (6)[1]	10	90-96 (6)	84	740
Deer	74-79 (6)	254	90-98 (7)	356	40
Squirrel	74-79 (6)	122	92-98 (3)	89	−27
Dove	74-79 (6)	150	92-98 (3)	109	−27
Quail	74-79 (6)	93	92-98 (4)	38	−59
Raccoon	79 (1)	25	90-92 (2)	16	−36
Rabbit	74-79 (6)	79	92-98 (3)	49	−38
Duck	74-79 (6)	24	92 (1)	16	−33

[1] Number of years in which hunters were estimated during the period in parentheses.

hunters and 66 percent of the total days of hunting were spent on deer.[13]

The large deer population is not without problems. Georgia has gone from cautious optimism about being able to increase the deer population at mid-century to moderate alarm about too many deer at the end. Odum and Turner (1987) stated the turn around as follows: "In fact, promotion of deer has been too successful since the one million deer now populating the state are beginning to have detrimental effects on crops, home gardens, auto accidents, etc. To avoid a "boom and bust" that is always inherent in exponential growth draconian measures (increased hunting, large scale removal, etc.) will now have to be implemented in order for this "run away" growth to be halted and some kind of desirable equilibrium established." The Georgia Department of Natural Resources (2000b) implemented, not draconian, but moderate measures to control the population after it peaked at about 1.3 to 1.5 million about 1990. The Wildlife Resources Division used increasingly liberal hunting regulations in an attempt to reduce the population to about one million. In 2000, the limit was up to eight deer per hunter; only two of those could be antlered bucks. It has been said that the deer herd is greater at the end of the twentieth century than when the white man came to Georgia. It could very well be true. If one assumes that there are one million deer in Georgia (there are likely more; the numbers in Figure 6.1A are referred to in the wildlife survey reports as minimum estimates), then the average number per square mile is about seventeen. This compares to an estimate that deer populations before Europeans came were as low as two per square mile over large sections of the mature deciduous forest in the United States.[14]

The "run away" growth has made the deer a nuisance to farmers and suburbanites in neighborhoods across Georgia. One of the more bizarre stories about deer in suburban areas was the entrance of an eight-point buck into a Kroger supermarket in Ellenwood, south of Atlanta in Henry County in February of 2001. The deer entered through an automatic sliding-glass door and damaged shelves and produce before being wrestled to the floor by the manager and some customers.[15] Even in urban Atlanta deer appeared in the 1990s where they had not been seen in 100 years. As Eldredge reported in 1999, "About 161 years after a buck's head was posted near present-day Peachtree and Roswell roads giving Buckhead its name, history has repeated itself. Twice." According to police reports, car-deer collisions occurred on Peachtree Road on consecutive nights. In 1998, there were 346 deer-car collisions in Henry County and 2,152 in the Atlanta metropolitan area.[16] There were injuries to 146 people in metropolitan area collisions; presumably, deer were injured or killed in all of them. Nick Nicholson of the Wildlife Resources Division of the Department of Natural Resources was able to document in 1998, 49,818 accidents between deer and vehicles using an index he created.[17] That's more deer accidents than the number of deer existing before the 1960s. The number may to be too high, because it means that a car hits one out of every twenty or twenty-five deer. But if it is reasonably accurate, at $2,000 per accident deer are costing Georgia motorists nearly one billion dollars per year.

Most of the deer that are killed by hunters in Georgia are dressed and put into the family freezer for later consumption. If the antler is unusually big, it may be mounted on the wall. One use of the venison that most people don't know about is its contribution to programs to feed the hungry.[18] From 1993 to 1998, hunters donated more than 16.5 tons of ground venison to feed thousands of Georgians during the holiday season. Inmates at the state prison at Alto process the deer, pack it, and label it with the "Hunters for the Hungry" logo. The venison is distributed to food banks located in Albany, Athens, Atlanta, Augusta, Columbus, Macon, and Savannah.

Wild Turkey

Wild turkey was one of the most plentiful game birds in eastern North America, so plentiful that Benjamin Franklin urged its adoption as the national symbol instead of the bald eagle. This bird must have been plentiful in Georgia when the settlers came, but clearing of the land and over-hunting decimated the population. Wild turkey numbers in Georgia reached their lowest between 1900 and the 1930s. According to Jenkins (1953) wild turkeys in South Georgia never were eliminated, and were quite plentiful on game preserves. However, "Only small remnant populations are found north of the fall line." In 1951, Georgia was estimated to have only about 22,000 wild turkeys, and sixty-three counties, mostly in the Piedmont, had less than ten turkeys each. There were about 500 turkeys in the mountain province, and according to Jenkins, the number was not increasing "sufficiently" in the 1950s. The greatest populations were in the Coastal Plain and the counties near the coast.

The primary causes for low numbers of turkey were said to be lack of habitat due to uniform growth of forest without sufficient clearings for growth of food, lack of burning to provide small open glades, illegal hunting, dogs, and possibly predation by foxes and bobcats.[19] Stoddard (1963b) thought that in Southwest Georgia there was not enough forest early in the century for wild turkey. He said that that in the twenty-five years preceding the 1960s, "The turkeys have built up fast with the buildup of timber in the Albany section." But he considered over-hunting the main cause of the earlier decline and continued low populations, saying that "The future of the Wild Turkey hinges largely on whether such men [poachers] can or will be controlled.... In some areas this lawlessness and coinciding turkey decline continue at this writing, with little indication what the final outcome will be."

Jimmy Carter tells about being surprised to see a wild turkey fly across the highway in front of him while driving near Savannah between 1966 and 1970.[20] He was apparently apprehensive about restoration of the wild turkey in Georgia, because he states, "I wondered if my baby daughter, Amy, would ever have a chance to witness such a sight before the noble creatures were all gone from our area of the world." He also relates his efforts at restoring the wild turkey to its former territory while governor. Thackston et al (1991) reported that the population had recovered to 17,000 in 1973. They apparently concluded the 1951 estimate (22,000 birds) given by Jenkins (1953) was too high or that turkey numbers had dropped further since the 1950s. The increase after 1973 was attributed to improved habitat and also to restocking which was intensified about that time. By 1991, the wild turkey population had increased to 375,000,[21] and in 2000 the estimate was 400,000 (Figure 6.2A).[22] According to Thackston et al (1991), "Wild turkeys now occupy 90% of Georgia's suitable habitat. The turkey population will continue to grow in many of these areas." Sweat (1992) called restocking of the wild turkey "One of the greatest success stories in modern wildlife management." The harvesting of wild turkeys by hunters has increased along with the populations. Surveys by the Georgia Department of Natural Resources (1993) show that the number of turkeys bagged increased from about 2,000 in the 1970s to 60,000 in the mid-1990s (Figure 6.2B). It is remarkable that, as in the case of deer, turkey hunters were harvesting more turkeys per year in the 1990s than existed in Georgia in the first half of the century. The number of turkey hunters increased a little more slowly, so that the turkeys killed rose from about 30 to just over 60 per 100 hunters.

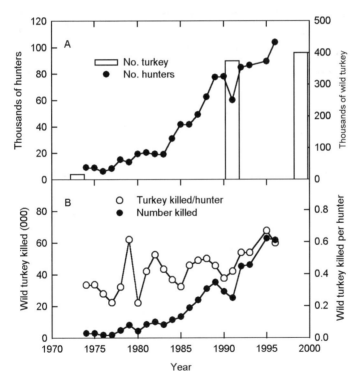

Figure 6.2. Changes in the numbers of wild turkey and hunters (A) and the number of turkeys killed and turkeys killed per hunter (B). From Thackston et al. (1991) and the Georgia Department of Natural Resources (1993, 2000c).

Quail and Dove

For much of the twentieth century quail and dove were the most popular and plentiful game birds in Georgia. Both are included in what was commonly referred to as "farm game" because their favored habitat is the fields and forest edges so prevalent on farms early in the century. Quail hunting was a popular pastime for the farmer and tended to be a sport for individuals or pairs of hunters. The hunt was often two or three hours stolen from farm work early or late in the day, and the birds were not hard to locate, because the farmer knew the general area of every covey on his farm. Many who grew up in the country never forgot the excitement of a birddog suddenly still in a "point," the explosive flutter of a flushed covey, and the quick shots necessary before the quail were out of range. Quail was a favorite meal enjoyed by farm families when the hunt was successful.

Dove shoots were often a community affair with farmers and townsfolk gathering at a field that was known to have plenty of feed and plenty of doves. According to Madson (1978) opening week of the dove hunting season in the southeastern states "is a blend of family reunion, lodge picnic, an old-style barbecue, and a Juarez election." A common location of the shoot in Georgia was a large cornfield "hogged off" by turning hogs in and allowing them to harvest the corn. Their inefficient harvest not only left considerable amounts of grain exposed in the field, but much of it was shelled and chewed, reducing it to a size easily consumed by the birds. Under cover of trees or fencerows, the hunters would wait for doves to flock to the field and fire away with birdshot. I remember as a boy, tales of doves coming so thick and fast that the shotguns would heat up to the point they could hardly be held. Doves were also a popular meal for the family when most Georgians still lived on the farm.

Where I grew up dove and quail were about equal in popularity, and the title of "most popular game bird" has been claimed for both. Baskett and Sayer (1993) claim that "The mourning dove is the leading game bird species in North America, with numbers harvested easily exceeding all other migratory game birds combined." Quail hunting has been much more recognized for its sporting qualities, requiring active hunting, cooperation between dog and hunter, and a climactic moment of quick and precise shooting. In many ways, quail has been typified as the game bird of fields and meadows. A "Cooperative Quail Investigation" was begun near Thomasville, Georgia in 1924 to improve the management of this bird and increase populations.[23] Quail hunting in Georgia has also been glamorized somewhat because the rich and famous hunted quail on large plantations in South Georgia throughout much of the twentieth century. At the end of the twentieth century quail was selected as the species to represent wildlife on license plates sold to raise funds for research and management.

Quail were probably not so plentiful when the European settlers arrived in Georgia. However, when forest was cleared and planted to crops, good nesting and feeding habitat was created. The accompanying fencerows and especially abandoned fields with broomsedge, briars, and weeds provided good cover and an abundance of food. Stoddard (1931) reckoned that soon after settlement, "With enormously increased food supply, and with lessened natural enemies, the bird in this early stage of agriculture experienced favorable conditions that perhaps never before or since have been equalled." Then for most of the nineteenth and twentieth centuries the patchy field and forest landscape was nearly ideal for quail. Not much is known about the actual numbers of quail in Georgia prior to the middle of the twentieth century, but at that time the population appeared to be quite variable over the state. The number ranged from fewer than 5 birds per 100 acres in much of the mountain region to 25–30 birds per 100 acres on small tracts of mountain farms and cleared valleys.[24] The more agricultural areas of the Piedmont and the upper Coastal Plain had 10–15 quail per 100 acres. Good quail populations were also found in parts of the Coastal Plain where pineland was burned frequently. Stoddard concluded (1939) that optimum pineland habitat for quail and maximum production of timber are not compatible, because maximum timber production requires thick stands of pines. Numbers of quail can be high for the first two or three years after clear-cutting and replanting of timber[25] if there are quail nearby to populate cut-over timberland. Coveys disappear, though, when trees grow large enough to shade out most of the wild legumes, grasses, and other small weedy plants that furnish food for quail.

The bobwhite quail numbers suffered major decreases during the second half of the twentieth century. Hardie (1999) stated, "Everyone agrees the quail population is plummeting. Their numbers here have dropped 70 percent in the past thirty years, according to the Department of Natural Resources. Consequently quail hunters also have decreased, from 135,000 to 42,000." The change in numbers of quail hunters in Figure 6.3A is not quite the same as given by Hardie, but both the number of hunters and quail bagged have decreased. In the state Wildlife Resources Division surveys there were approximately 115,000 hunters in 1974 and that number dropped to about 40,000 in the 1990s. The number of birds harvested decreased to nearly the same extent so that birds harvested per hunter each year did not show a definite trend. In surveys by the US Fish and Wildlife Service, the number of quail hunters in Georgia was 150,100 in 1985,[26] but only 57,000 in 1996.[27] These numbers are higher than those estimated by Georgia game management personnel (Figure 6.3A), but the decrease is just as steep. Quail hunters made up 28 percent of total hunters in 1985, but only 14 percent in 1996.

Decreases in the number of quail hunters may be caused by decreases in the number of hunters who live in rural areas and increased populations and popularity of larger game. But the change that causes the most concern is the decrease in quail numbers and habitat. Most reports of the decline in quail numbers cite opinions by hunters or game experts, but the few counts that have been made also show this decline. The North American Breeding Bird Survey shows that the number of bobwhite quail in Georgia has decreased by an average of 4.3 percent per year since 1966 (Table 6.2).[28] Another survey, the National Audubon Society Christmas Bird Count, resulted in estimates of 5.4 percent decrease per year from 1959 to 1988.[29] There has been enough recent concern about the decline that during the 1999 session the Georgia General Assembly

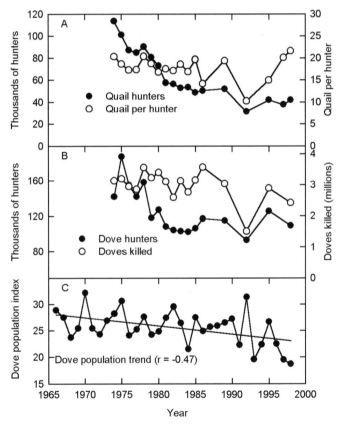

Figure 6.3. Changes in number of quail hunters and quail bagged per hunter (A), number of dove hunters and doves killed (B) and the population index of doves in Call Count Surveys (C). From the Georgia Department of Natural Resources (1993, 1993–1999) and Dolton and Smith (1998).

Table 6.2. Bird species that decreased or increased in Georgia from 1966 to 2000 (North American Breeding Bird Survey - Sauer et al., 2001) and from 1959 to 1988 (Christmas Bird Count - Sauer et al., (1996). Only those trends indicated as statistically significant (<0.1 probability) are included.

| | % / year | | | % / year | |
| | BBS[1] | CBC[1] | | BBS | CBC |
Decreasing species	1966-2000	1959-88	Increasing species	1966-2000	1959-88
House sparrow	−6.7	−3.4	Carolina wren	1.0	NC[2]
Grasshopper sparrow[3]	−5.0	—[2]	Blue grosbeak[3]	1.4	—[2]
Common nighthawk[3]	−4.9	—	Grt. crested flycatcher[3]	1.4	—
Hairy woodpecker	−4.4	−1.3	American crow	1.5	4.6
Northern bobwhite	−4.2	−5.4	Black-and-white warbler[3]	2.2	—
Northern flicker	−4.0	−1.9	Killdeer	2.8	NC
Eastern meadowlark	−3.7	−5.4	Tufted titmouse	2.8	1.6
Rock dove (pigeon)	−3.3	—	Blue-gray gnatcatcher[3]	3.1	2.4
Yellow-billed cuckoo[3]	−3.0	—	Kentucky warbler[3]	3.3	—
Blue jay	−2.4	NC	Black vulture	3.3	−3.4
Common grackle	−2.2	NC	Yellow-breasted chat[3]	3.8	—
Red-winged blackbird	−2.3	NC	Song sparrow	3.9	−1.3
Field sparrow	−2.0	−4.4	Eastern bluebird	4.5	1.9
Northern mockingbird	−1.9	NC	N. Rough-winged swallow[3]	4.8	—
Eastern towhee	−1.5	—	Great horned owl	5.0	NC
Mourning dove	−1.2	3.2	Yellow-throated vireo[3]	5.3	—
"	−0.5[4]	—	Barn swallow[3]	5.6	—
Chimney swift[3]	−1.1	—	Turkey vulture	5.7	3.3
Northern cardinal	−0.9	NC	Great blue heron	6.0	2.2
Rusty blackbird	—	−16.7	Red-shouldered hawk	7.5	NC
Slate-colored junco	—	−5.3	Eastern phoebe	8.3	1.1
Rufous-sided towhee	—	−3.7	Barred owl	13.3	—
Evening grosbeak	—	−3.5	Canada goose	23.2	—
Fox sparrow	—	−3.0	House finch	30.7	—
Brown thrasher	NC	−2.7	Redhead	—	1.7
Wh.-throated sparrow	—	−2.6	Belted kingfisher	NC	1.9
Northern pintail	—	−2.3	Red-breasted merganser	—	1.9
Brown creeper	—	−1.6	Pileated woodpecker	NC	2.1
			Pine warbler	NC	2.5
			Screech owl	—	2.7
			Carolina chickadee	NC	3.2
			White-eyed vireo[3]	NC	3.6
			Red-tailed hawk	NC	3.7
			Solitary vireo	—	3.8
			Cedar waxwing	—	4.0

[1] BBS = North American Breeding Bird Survey; CBC = Christmas Bird Count. [2] NC = no change, — = not included in analysis. [3] Species classified by Sauer et al (2001) as neotropical migrants. [4] Estimate from Call Count Survey (Dolton and Smith, 1998).

appropriated $939,000 to boost the bobwhite quail population, and passed legislation to provide future funding.[30] The Georgia Department of Natural Resources (2000d) instituted a program called the Bobwhite Quail Initiative in which landowners agree to establish and maintain hedgerows and field borders as "early successional habitat." The program was started in seventeen counties in Central and Southwest Georgia and the goal is to establish a total of 120 miles of hedgerows and field borders suitable for quail. Altogether, about 100,000 acres are involved and landowners are paid on a per-acre basis for land established as quail habitat.

The concern about decreasing quail is not just a recent one. According to Landers and Mueller (1986), "As quail numbers declined in areas protected from fire, a group of plantation owners in South Georgia and North Florida formed the Cooperative Quail Investigation in 1924. Their project leader, Herbert L. Stoddard, was directed to find the cause of the problem as well as its solution." The solution was not a quick and easy one because an editor of the *Albany Journal* called in 1940 for a three-year "holiday" for quail because of what he described as "fast diminishing quail life."[31] Landers and Mueller (1986) discussed the variation of quail numbers due to weather, predators, habitat, and hunting pressure and concluded that they are somewhat cyclic, making the demonstration of trends difficult. Even though early decreases in quail populations were thought to be mostly a result of the decreased burning of woods, later in the century the main cause was conversion of cultivated and idle land to productive forests. So the successful twentieth-century campaign to protect and expand Georgia's forest was one of the main reasons for the decline of quail.

The mourning dove is the most abundant and widespread game bird in the United States. In a 1989 survey, doves ranked second in number of all birds counted only to red-winged blackbirds. The widespread clearing and farming of large areas of the United States increased dove habitat and numbers "so that they are probably more numerous now than in colonial times."[32] Unlike the bobwhite quail, doves are migratory birds; those in the northern latitudes migrate great distances.[33] Because doves depend on similar habitat as quail (open areas, grain fields, and forest edges), their numbers might also have been expected to decline. There was a decrease, smaller than for quail, but statistically significant, in two separate sets of surveys conducted since the 1960s. Federal and state biologists conduct "Call Count Surveys" across the United States for the US Fish and Wildlife Service. These surveys are done on 20-mile routes, mostly on secondary roads where dove calls are counted at 1-mile intervals. The second set of surveys is the North American Breeding Bird Survey described later in the section on songbirds. In the Call Count Survey, the decline in doves is estimated to be 0.5 percent per year from 1966 to 1998 (Table 6.2, Figure 6.3C),[34] but in the Breeding Bird Survey[35] the decrease is estimated to be twice as great (Table 6.2). The Breeding Bird Survey was considered by Martin and Sauer (1993) to give more reliable estimates of trends than the Call Count Survey because of the larger number of routes used. However, in a third survey, the Christmas Bird Count conducted by the National Audubon Society, the dove actually increased in Georgia at 3.2 percent per year from 1959 to 1988.[36] The increase at Macon was spectacular, from 15 yearly sightings per year before 1950 to over 400 in the 1990s. This survey is made at Christmas time each year and involves rural and urban areas. The increase in this survey may mean that more doves are wintering in Georgia than earlier in the century.

In a game survey in the middle of the twentieth century, hunters were of the opinion that there was an alarming decline in dove populations in the previous twenty years.[37] Whatever the decrease before the 1960s, the later changes were fairly minor as indicated above, and the number of doves killed remained nearly steady from 1974 to the early 1990s (Figure 6.3B). In 1992, the number dropped to nearly 1.5 million, but later in the 1990s rebounded to between 2 and 3 million. According to Jenkins, the perception of decreased numbers near mid-century may have been caused by the spreading of doves due to more widespread planting of crops (lespedeza, small grain, and pasture) that they favor. But a decrease prior to the 1960s may have also have been real, because Jenkins (1953) estimated an annual kill of two million at mid-century, one-third lower than the average of 3.12 million per year killed from 1974 to 1989 (Figure 6.3B).[38]

Overall, there appears to have been a decrease in doves since the middle of the twentieth century (Figure 6.3B), but the decrease may depend on the location and time of year. The decrease has apparently been small and the dove is still plentiful. Reeves and McCabe (1993) attributed the high numbers to the adaptability of the dove, saying, "The mourning dove has adapted to—indeed usually prospered from—human uses and abuses of land." The dove is so adaptable that it not only nests in forests and feeds in fields, but it nests in suburban neighborhoods and eats at backyard feeders. More Georgians today see doves in their backyards than in open fields, because both the birds and the people have changed. The decrease in populations shown in Figure 6.3C should not be of particular concern for a species as adaptable as the mourning dove. In fact, the decreasing trend (if it is real) is probably an inevitable result of the shift of the landscape from small farms to large continuous tracts of forest.

The number of dove hunters has decreased from about 175,000 in the 1970s to about 100,000 in the 1990s (Figure 6.3B, Table 6.1). The decrease was about 30 percent over that period compared to about 60 percent for quail hunters. Although the numbers vary, it is fairly clear that dove hunters in Georgia have decreased more than doves. The Georgia Department of Natural Resources has made an effort to increase doves and dove hunters by managing dove fields totaling 1,171 acres on twenty-eight wildlife management areas throughout the state.[39]

Rabbits and Squirrels

In the early 1950s when there was still an abundance of row-crop agriculture, fencerows, and abandoned fields, the number of rabbits and interest in hunting them were high. Jenkins (1953) evaluated the situation this way; "Rabbit hunting is very good over much of Georgia and it is gaining in popularity due largely to its general availability and the excellent table qualities of rabbits." "Excellent table qualities" will be a strange characterization of rabbits for most modern Georgians, who have never eaten rabbit. The popularity of rabbit hunting and eating was nationwide earlier in the twentieth century, as shown by a survey by the Remington Arms Company.[40] They found that rabbit hunters used about 30 percent of the shotgun shells they sold. The next most popular use of their shells was by squirrel and quail hunters, who used about 14 percent each.

Popularity of rabbit hunting waned, however, late in the twentieth century, dropping by 38 percent in Georgia from the 1970s to the 1990s (Table 6.1). In fact, in a 1996 survey of hunting and fishing by the US Department of the Interior and US Department of Commerce (1998) the number of rabbit hunters in Georgia was not even listed. The number of rabbits killed in Georgia decreased by about the same percentage as the number of hunters. Because the number of rabbits killed per hunter did not decrease over the last two and one half decades, a scarcity of rabbits was not likely the cause of the decrease in rabbit hunting. In fact, it is not known if the rabbit population decreased late in the twentieth century, but its favored habitat, fencerows, broomsedge, and briar patches has. This was especially true for the Piedmont where rabbit hunting was so popular. In addition, coyotes were introduced into Georgia and became quite numerous by the end of the century. Rabbit is a favorite prey of the coyote.

Early in the twentieth century squirrels were nearly as popular with hunters as rabbits. Most farms had some woods with enough food to attract and maintain squirrels. In mid-century Jenkins (1953) reported that squirrel numbers appeared to be rather low over much of north Georgia and only moderate over the Piedmont. Excellent populations were found over most of the Coastal Plain. Production of acorns, nuts and other squirrel food was considered to be the main factor in determining their populations in Georgia. Hunting of squirrels has declined in much the same way as rabbits. As shown in Table 6.1, squirrel hunters decreased by 27 percent in the last three decades of the twentieth century. Squirrel hunters were estimated at 86,000 in Georgia in 1996,[41] but the report noted that estimates for small game hunters were based on fewer than 30 responses out of about 465 people interviewed and therefore may not be accurate. As in the case of rabbit, the decrease in squirrel numbers is not likely the reason for decreased hunting. An increase in the abundance and popularity of larger game is a more likely reason. Although numbers of squirrels are not estimated, suburbanites know that they are plentiful in the neighborhoods, perhaps even more plentiful than in the forests.

Hunting of all small game has decreased in Georgia since the mid-twentieth century (Table 6.1). The change is reflected mainly in the shift from small to large game. Out of 476,000 hunters in 1980, 59 percent hunted small game (some, of course, hunted both). Five years later small game hunters had dropped to 54 percent, and in 1996 only 31 percent of 403,000 hunters sought small game. It is risky to add averages such as those in Table 6.1 because they represent different years and unknown variation. Nevertheless, the sum of the average numbers of hunters are the same for the 1970s and the 1990s, indicating that hunting has not decreased much, but has shifted from small game to wild turkey and deer. The main reasons for decreased hunting of small game are likely the decreased rural farm population, the decrease in use of small game for food, and the increased availability of other, larger game. The number of hunters who lived in rural areas of Georgia decreased by

(A)

Figure 6.4. A beaver dam (A) on Long Creek in Oglethorpe County and a portion of the pond where trees have

half from 1985 to 1996.[42] When rabbit and squirrel were the most popular game animals they were hunted largely for food. These species have no trophy animals and few current Georgians would eat them, much less hunt them for food. An increase in affluence of hunters and the attractiveness of larger game are likely reasons that small game is not as popular as it used to be. Of the $713 million spent by Georgia hunters in 1996 for equipment, supplies, etc., only 6 percent was spent on small game.[43] The decline in hunting of small game is not peculiar to Georgia; hunters in this category decreased nationwide by one-half from 1975 to 1991.[44]

Furbearers

Before Georgia was settled, beaver was plentiful. They probably supplied the Indians with both meat and hides. Wharton (1977) suspected that Indians drained beaver ponds to use as cornfields because of the accumulation of fertile sediment in the ponds. The beaver was almost eliminated, however, because their habitat was destroyed, especially in the Piedmont where lands were cleared in many cases down to the stream banks, and also because their skins

been killed by flooding (B). The dam is about 200 yards downstream from the upper area shown in (B).

were valuable. Georgia was essentially devoid of beaver early in the twentieth century until restocking efforts began in the 1930s.[45] About their scarcity Jenkins (1953) said, "As recently as ten years ago one author considered beaver in Georgia as practically extinct." At about the same time Chapman et al (1950) remarked "Once common, the beaver is now rare."

They were not extinct and not very rare in some parts of the state in the 1950s, however, because in South Georgia there were trapping seasons at the insistence of local timber owners. When beaver built dams in flat valleys they flooded out several acres of pineland.[46] All but three Georgia counties had reported beaver damage to timber in 1975, and "Between 1967 and 1975, beaver damage increased 128%, with 287,700 Georgia acres inundated."[47] The Georgia Senate passed a resolution in 1977 urging the Department of Natural Resources to make a study to determine methods for more effective control of beavers. Figure 6.4 shows two views of a typical beaver pond in the Piedmont. The dam extends not only across the channel, but across the stream valley flooding several acres and killing most of the trees.

The beaver has been a destructive nuisance or a beneficial conservationist, depending on the bias of the observer. Chapman et al (1950) was complimentary of beavers for their dam building, saying "Beavers aid in erosion control by catching runoff waters in their ponds. Their dams improve fishing by creating pools." He was hopeful that "Careful management may yet bring this industrial and useful animal back to our favorite streams." The numbers had increased enough in the 1970s that Hicks (1977) concluded that "Beaver ponds are largely responsible for the currently high population of wood ducks..." Diversity and abundance of fish were both increased in areas affected by beaver. Beaver ponds have been found to contain more fish of a size large enough to be of interest to sport fishermen than their feeder streams.[48] But the flourishing beaver did not only affect fish, ducks, and timber. At the end of the twentieth century, the beaver had moved into the suburbs, prompting calls from homeowners for trapping and removal of the dam-builders.[49] Whether the beaver is considered a blessing or a pest, its numbers have increased greatly in the twentieth century.

The bobcat was feared as a nuisance and predator when small farms covered the landscape. They were scarce in Georgia early in the twentieth century, but never disappeared from the state. According to Jenkins (1953), low populations prevailed over most of the mountainous region in the early 1950s and bobcats were present but rare in the Piedmont. However, they were common to abundant over much of the Coastal Plain, especially in river swamps where they were not much danger to livestock. A statewide survey of 210 "scent stations" was conducted beginning in 1978.[50] The stations were prepared to attract animals and visits were recorded. A relative abundance was calculated from the visits as follows: Relative abundance = (Total animal visits/Total operative station nights) X 1,000.

The statewide relative abundance for bobcat decreased by about half from 1978 to the 1980s and increased again (Figure 6.5A).[51]

The number of bobcat trapped increased from 2,772 in the 1977–1978 season to about 4,000 for the next three years and then decreased to about 600 per year in the early 1990s (Figure 6.5B). For the rest of the 1990s the numbers trapped were about 1,000 to 1,500. A study of harvested bobcats from 1977 to 1980 showed that it was made up of mostly juvenile and yearling classes, indicative of heavy harvest pressure.[52] The shooting of bobcat was prohibited beginning in 1984, and in later years juveniles and yearlings made up about half or less of the harvest. Although abundance estimates are not available after 1992, the decrease in trapping apparently results from causes other than abundance of animals, which was higher in 1992 than in 1980. This conclusion is strengthened by the

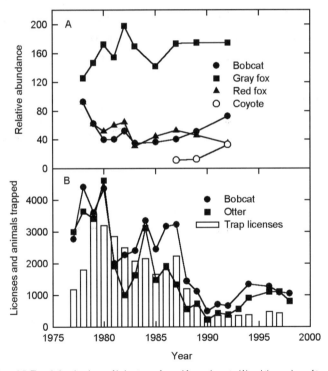

Figure 6.5. The relative abundance of bobcat, gray fox, red fox, and coyote (A) and the numbers of trapping licenses sold and bobcat and otter trapped in Georgia (B). From the Georgia Department of Natural Resources (1993, 1993–1999).

fact that the number of trapping licenses issued in Georgia (total, not just for bobcat) followed the numbers of bobcat trapped, dropping from 3500 in 1979 to less than 500 per year in the 1990s (Figure 6.5B). The decrease in trapping of otter is also likely due to a decrease in trappers rather than a decrease in otters. The number of otters in Georgia has likely increased, because they were restricted to the Coastal Plain in the first half of the twentieth century, but have since spread to nearly every county.[53]

Fox hunting was a popular sport in rural Georgia in mid-twentieth century and earlier. The gray fox is the more numerous species and the most hunted, but the red fox is more challenging, especially in open country, because it is less likely to be caught or cornered. Although fox is usually thought of as an animal of open country, their numbers do not appear to have suffered from the landscape changes of the last half-century. Swiderek reported relative abundance values for the gray fox at between 120 and 200 for the period from 1978 to 1992 with some indication of increase in the 1970s and early 1980s (Figure 6.5A).[54] Red fox populations, on the other hand, are half or less compared to the gray fox and decreased 60 percent from 1978 to 1992. Swiderek attributed the decline in red fox to the increase in coyotes because they are reported to kill or displace red fox. However, the number of coyote did not appear to be very high during the 1970s and 1980s when red fox populations were decreasing the most (Figure 6.5A).

The raccoon is a common animal in Georgia, and although hunted and trapped for its fur quite heavily in times past, its numbers seem to have remained high. In the surveys done by Swiderick, relative abundance in Georgia remained nearly steady at about 70 to 80 from 1980 to 1992. These surveys showed a much higher relative abundance in South Georgia (100) than in the mountains (9), with abundance in the Piedmont in between. According to Swiderick, raccoon and gray fox show fluctuations in numbers that correspond to each other, probably reflecting the susceptibility of both species to canine distemper. Distemper and rabies tend to cause population crashes, which are followed by rapid build up.

In 1937, Georgia was estimated to have 208 black bear.[55] Today there are about ten times that many mostly in the north Georgia mountains, along the Ocmulgee and Altamaha River drainage systems, and in the Okefenokee Swamp.[56] Bear in the North Georgia mountains have become more numerous in recent years and occasionally come in to the northern suburbs of Atlanta. In May 2000, five black bears were seen in subdivisions of Sandy Springs and Cumming and on a golf course in Cobb County.[57] Visitation of about 550 sites where biologists keep track of black bear populations in eleven north Georgia counties increased from 19.0 percent (within five days) in 1987 to 32.3 percent in 1993.[58] The bears of South Georgia have also become numerous enough to sometimes escape the swamps. In February 2001 a 350-pound black bear wandered into a hotel-restaurant district of Tifton.[59]

Alligator

Alligator numbers have grown from very low to nuisance levels in some areas of Georgia.[60] After populations dropped to a low point in the 1960s, the alligator was listed as "endangered" by the Georgia Department of Natural Resources.[61] Total numbers of alligators for the state increased from 29,954 in 1973 to 101,644 in 1982, but the Department of Natural Resources considered the state-wide estimates so uncertain that they were discontinued that year.[62] By 1984, surveys in several locations indicated that the population had increased to the point that endangered status was no longer necessary, and the alligator was removed from the state's Endangered Species List.

The alligator has gone from endangered to endangering. From 1980 to 1988 there were 1,954 complaints to the Department of Natural Resources about nuisance alligators. Nearly half of the complaints were about loss of pets and livestock or concern for the safety of animals as well as humans. About 1,100 of these alligators were captured and relocated. Over 50 percent of these complaints came from eight counties along the Georgia coast. Chatham and Glynn counties accounted for 32 percent of all complaints in the state. The number of complaints reached a high of 657 in 1991, and decreased to 386 in 1992. It is uncertain why the number dropped drastically from 1991 to 1992, but high rainfall and high water levels during the later count may have allowed alligators to remain in more remote areas.[63]

Waterfowl

Wild Canada geese were virtually unknown in most of Georgia in the middle of the twentieth century, except by fly-overs in their migrations between north and south. They occasionally landed, but seldom stayed more than a day or two in locations that provided water for a rest. Now they are permanent residents of the state in large numbers. The goose flock at Lake Walter F. George increased 200 percent from 1969 to 1976.[64] Between 1976 and 1987, over 8,000 Canada geese were relocated to Georgia from Tennessee and northeastern states.[65] The resident population has increased steadily since the mid-1970s when there were nearly none to about 45,000 at the end of the twentieth century (Figure 6.6).[66]

Although still relatively small, the wild goose population has become large enough to be a nuisance in some locations. The Georgia Department of Natural Resources (1993) started a project to "Control the size and/or growth rate of resident Canada goose flocks in areas where such flocks are creating nuisance or other problems." To thin the population an experimental goose-hunting season was tried for three years (1990–1992) in six units comprising a total of fifteen counties in the Piedmont region. There was said to be pressure to extend the seasons in some areas because of the increase of nuisance complaints. The population of geese on Lake Seminole in Southwest Georgia has increased to about 700 in the early 1990s and "is beginning to stimulate more nuisance problems."[67] The nuisance nature of increased geese populations is nationwide as a news release of the US Fish and Wildlife Service shows.[68] Most Canada geese nest in the Canadian Arctic, but according to the news release, "...increasing urban and suburban development in the U. S. has resulted in the creation of ideal goose habitat conditions—park-like open areas with short grass areas adjacent to small bodies of water." The result of restocking and creation of this habitat is that geese are breeding near small ponds of parks and golf courses, and other lakes and ponds all over Georgia. More than a nuisance is the probability that geese contribute to the contamination of recreational areas. Geese have been suspected of fecal coliform contamination of areas on Lake Lanier[69] and a public beach on Lake Acworth.[70]

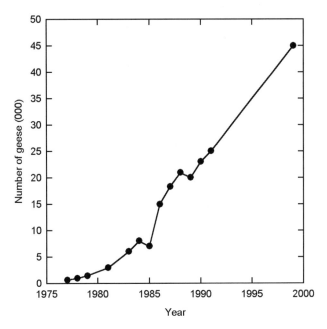

Figure 6.6. Changes in the number of resident Canada geese in Georgia from 1977 to 1999. From the Georgia Department of Natural Resources (1993, 1993–1999).

Changes in the abundance of ducks in Georgia are difficult to assess because there are so many species that fall in this category and they have not been tallied very well. Hicks (1977) credited the beaver with the comeback of Georgia's most common species, the wood duck, "a species which a few decades earlier, was practically extinct." "Practically extinct" may not be a very accurate description of its status, because Greene et al (1945) listed the wood duck as fairly common along the rivers and swamps throughout the state. Almand (1965) noted that in the early years of the Piedmont National Wildlife Refuge (1930s) only one or two ducks would be seen along Falling Creek or the nearby Ocmulgee River. As soon as man and beaver began building ponds on the refuge, populations increased. Three small man-made water-holding structures were built in the late 1950s and numbers more than doubled from 1953 to 1959. In 1964, before the winter migration into the area, there were about 240 wood ducks.

In the Christmas Bird Counts in the 1990s there were an average of 450 wood ducks on the Piedmont National Wildlife Refuge. What is more remarkable, there was an average of 1,535 ring-necked ducks and nearly 3,000 ducks of all kinds. The increase in ducks was not limited to the wildlife refuges. In Macon, the number of ducks in the Christmas Bird Counts increased from an average of 6 in the 1940s and 1950s to over 900 in the 1990s. There are also high numbers of the mallard, ring-necked, and merganser ducks in Christmas Bird Counts at other locations across Georgia in the 1990s. The numbers of winter migrant ducks have increased, but there are few records of increases in nesting of ducks in Georgia. Regardless of the records, ducks are fairly easily spotted on ponds across the state, especially those in parks and other areas where picnickers supply food. Although the number of ducks appears to have increased in Georgia there is no evidence that hunting has increased; it appears to have decreased according to the Georgia Department of Natural Resources (1993, 1993–1999) (Table 6.1). The US Fish and Wildlife service estimated that there were about 35,000 duck hunters in Georgia in 1985,[71] but the estimate is projected from a small sample size. In the 1990s, estimates have averaged about 10,000 hunters.[72]

Nongame Species

Since the beginning of public funding of wildlife research and management, most of the funds have been spent on game species. Late in the twentieth century, however, increased attention has been paid to nongame animals and birds. Two popular initiatives used to raise money for nongame species research are the wildlife license plates sold for cars and the check-off of dollars on individual tax returns in Georgia. In 1997, the State of Georgia started selling car license plates featuring the bobwhite quail to raise funds for nongame wildlife research and management. The Georgia Department of Natural Resources announced in July 1999 that more than 650,000 license plates had been sold, raising more than $9 million.[73] For each license plate sold, $12.41 is used by the Wildlife Resources Division's Nongame Wildlife-Natural Heritage Section for conservation and management of nongame wildlife, some of which are endangered.[74] According to a press release of the Department of Natural Resources, the nongame research program receives no state funding for its projects. The income tax check-offs and the wildlife tag sales are the primary funding for projects on bald eagles, manatees, frogs, salamanders, loggerhead sea turtles, among the more than 950 species of nongame wildlife. In spite of the funds generated by these two programs and the prominence of many nongame species, most of the wildlife funding is still for game species. However, some of the most notable recent successes in wildlife management have been with nongame species.

The removal of the bald eagle and the peregrine falcon from the endangered species list within a few months of each other in the late 1990s was a tremendous achievement. The announcements were also two of the few recent instances of published good news about the environment. Populations of both birds have recovered in Georgia and the rest of the United States. In the lower forty-eight states, nesting pairs of bald eagles increased from 417 in 1963[75] to 5,748 in 1999,[76] an increase of 1,278 percent in thirty-five years. Frank Gill, editor of the book The Birds of North America said "They are now more common than they were at the turn of the century" and "This is one of the big successes of this century." Between the late 1960s and 1980 there were no eagle nests in Georgia, but in the 1999–2000 season, there were forty-seven successful nests in forty-one counties producing a total of seventy-eight young eagles.[77]

The peregrine falcon also recovered enough to be taken off the endangered list in 1998. Secretary of the Interior Bruce Babbitt selected Atlanta as the site of the announcement.[78] The secretary said he came to Georgia to announce the removal because of "the fine spirit of conservation in the state" and because of organizations that had helped the peregrine falcon recover in Georgia, including the Department of Natural Resources, Georgia Power Company, and the Chattahoochee Nature Center.

The wood stork is a species that once occurred in large numbers in South Florida, but those numbers decreased by 75 percent from the 1960s to the 1980s.[79] The US Fish and Wildlife Service classified the wood stork as an endangered species in 1984. The number of birds apparently stabilized or perhaps increased in the Southeast in the 1980s. Wood stork nests were first found in Georgia on the Blackbeard Island National Wildlife Refuge in McIntosh County in 1965. Its population in Georgia was apparently never very large, but has grown from very small in the 1960s to over 1,600 in 1993.[80] The Birdsville colony was discovered near Millen, Georgia in Jenkins County in 1980. The number of nests in that colony increased from 100 in 1980 to 330 in 1993. Because of the scarcity of the birds in mid-century and a remarkable increase since the

1980s, it is surprising that the North American Breeding Bird Survey counts show no change since 1980.[81] As expected for an endangered species, the wood stork was found in a small number of survey routes (sixteen compared to sixty-eight for quail and mourning dove, for example).

The osprey is a water bird that is not well known in Georgia. Denton et al (1977) describes it as an uncommon summer resident and rare in winter, nesting in large wetland areas such as coastal wetlands, the Okenofenokee National Wildlife Refuge, Lake Seminole, and in Pulaski County. It is an "uncommon visitor elsewhere." Riddleberger and Odom (1987) reported that ospreys increased in Georgia from an estimate of 50–55 pairs in 1977 to almost 200 pairs in 1987. They indicated that construction of large lakes in the twentieth century has caused some osprey to nest away from the coast. One nest was reported in Lumpkin County on Lake Lanier.[82]

One nongame species that Georgia could do without is the coyote. Coyote is not native to Georgia. This species was apparently brought into the state by private citizens as early as the 1920s,[83] perhaps to be hunted by foxhounds. So its reason for being in Georgia could be that it was considered to be "game." Coyote were first widely reported in Georgia in the 1950s. They presently inhabit all counties in the state, although in the mid-1980s they were concentrated in southwestern Georgia.[84] By the late 1980s their numbers had started to increase statewide (Figure 6.5A) and at the end of the century they had even moved into populated areas near the larger towns. Schrade (1999) reported that coyotes were killing turkeys on a farm in east Marietta and that the control tower at the Cobb County-McCollum Airport warns approaching pilots to be on the lookout for coyotes. The superintendent of Kennesaw Mountain National Battlefield Park related that coyotes have sought refuge in the park and other open areas of west Cobb County. Coyotes normally eat small animals such as rabbits, rats, chipmunks, birds, and lizards. When food is scarce, however, coyotes attack livestock and even pets. According to Bailey (1997), Georgia farmers lost 200 lambs to coyotes in 1994, and 100 head of cattle and 800 calves in 1995.

Songbirds

Georgians, like other Americans, appreciate songbirds, but the appreciation has undergone a wholesale change in the twentieth century. From nearly indiscriminate shooting early on, attitudes have evolved to social and legal prohibition of shooting of songbirds. The Christmas Bird Counts of the Audubon Magazine were started at the beginning of the century partially as an alternative to Christmas-time "side hunts" in which as many birds as possible were killed. John James Audubon is quoted as saying about robins in the 1800s, "Every gunner brings them home by the bagful."[85] As incredible as it sounds, the writer also said of Audubon himself, "He shot birds like mad, often far more than he needed for his studies." A measure of the change of attitudes toward songbirds was the custom, begun about the middle of the twentieth century, for towns all over the state to erect signs at the city limits declaring them "Bird Sanctuaries." This was a signal of an increasing awareness of the value of songbirds and a determination to protect them. An earlier signal was the selection of the brown thrasher, a songbird with no value as game, as the state bird. The shooting of birds such as robins and brown thrashers today would offend most Georgians. There are hundreds of thousands of bird feeders at homes where people enjoy watching birds come and go. In a 1996 survey in Georgia, it was estimated that 1.5 million Georgians feed some form of wildlife,[86] mostly at bird feeders.

During the height of environmental awareness after the publication of Silent Spring there was concern about demise of many songbirds due to their consumption of insects and other foods containing pesticides. Even at the end of the twentieth century there was concern about loss of songbirds. Kundell (1996) stated that "Populations of many songbirds in the South are declining due to a variety of reasons...." In a 2000 news release, the Georgia Department of Natural Resources (2000d) claimed that "Changes in farming and forestry practices, along with development and other land-use changes... consequently resulted in lower quail and songbird populations in Georgia." In 2001, Seabrook listed thirty-eight Georgia songbird species "in peril." Odum et al (1993) recounted the common theme about bird populations, "...there is much concern about declines in numbers in many species, the growing list of endangered species and the loss of biodiversity in general." However, they also acknowledged that "The other side of the picture, where ranges have been extended, or population size increased has received less attention." The following sections will focus on "the other side of the picture." Most of the birds discussed are songbirds, but a few do not sing, at least songs that humans care about.

The hundreds of species of small birds have different requirements and some species thrive in conditions that harm others. During the height of the cotton-tenant-farming period, many songbirds thrived in the patchwork of cotton, corn, fencerows, abandoned fields, and farm woodlands. Some, such as house wrens, purple martins, and pigeons are encouraged by, if not dependent on, domestic habitat. But other species prefer the forests. So the shifting landscape from forest to cultivated fields and back to forest and suburbs was bound to change bird life, decreasing some species and increasing others, confirming Leopold's view that "Wildlife is never destroyed...it is simply converted from one form to another."[87]

In contrast to the warnings about the disappearance of songbirds, fourteen species have extended their breeding range into the Athens area on the Georgia Piedmont since 1946.[88] Half of the new species live in what the authors called "domesticated habitats, while the rest are forest or forest-edge species. The extension of their breeding range was because of "urbanization, crop changes, and replacement of farmland by forest in rural areas." No species were found to have left the area since 1946, but several species of open areas were said to have declined. The movement of the new species into the area was described as "enriching a historically depauperate avifauna."

On a larger scale the decline of some birds in Georgia has been offset by increases in others. In the mid-1960s, the North American Breeding Bird Survey began using amateur bird watchers across the country to estimate bird populations. These "birders" drive designated routes and count birds they hear and see. Each route is 24.5 miles long, with a total of fifty stops located at half-mile intervals.[89] When the Georgia surveys were analyzed for ninety-four species, it was found that eighteen species had decreased from 1966 to 2000, but twenty-four had increased (Table 6.2). The remaining fifty species had not changed. Although some authors have listed all species as either declining or increasing, only those with statistically significant changes are listed in Table 6.2. There are many more than ninety-four bird species in Georgia, but only those encountered in at least fourteen routes over the years were included.

The increased or decreased numbers of some birds may reflect a change in their habitat. Of thirty-three woodland birds, eight increased from 1966 to 2000, two decreased (the yellow-billed cuckoo and hairy woodpecker), and twenty-three remained unchanged.[90] Only two grassland birds (the eastern meadowlark and grasshopper sparrow) were included in the analysis and both decreased probably because there are fewer open grassy areas in Georgia than earlier. Out of twelve "urban birds" seven decreased and only one increased, the house finch. The mourning dove was one of the declining species. It is somewhat surprising that this bird is classified as urban, because it flocks to grain fields in great droves and breeds in forest edges near open fields.[91] Many suburbanites will know, however, that doves nest in their neighborhoods.

Although not listed in the survey as an urban bird, the American crow was listed as increasing in Georgia since 1966, and as many suburban residents will testify, it has become much more urbanized. The classification as "urban birds" must be somewhat arbitrary. The purple martin was a favorite bird on many small farms in the middle of the twentieth century where "martin houses," gourds, or other nesting containers were hung to attract these birds. Sauer et al (2001), however, classified them as urban birds, and they were unchanged in numbers from 1966 to 2000. A Department of Natural Resources (2000e) press release states that the purple martin has come to rely almost entirely on manmade nesting places. Except for the decrease in seven "urban birds," the above changes probably result largely from the shift from row-crop agriculture to forests and non-agricultural lands.

It is not clear why urban birds should decrease in Georgia as the state has become more urbanized, but most of the ones that have declined are well-known to Georgians. For some reason pigeon (rock dove) numbers have declined (Table 6.2). The reason for its decline is not likely loss of habitat, which is widely blamed for decreases in population of its relative, the mourning dove, because the pigeon nests and roosts in buildings and other structures. Programs to eradicate or discourage pigeons from urban areas or decreases in food supply may be responsible for the decline. The other "urban" birds that have declined are the mockingbird, house sparrow, blue jay, grackle, and chimney swift.

In addition to the Breeding Bird Survey the National Audubon Society began in 1900 counting birds across the country at Christmas time. This survey is called the Christmas Bird Count and it has been conducted in Georgia for many years, in some places back to the early 1900s. In this survey all of the birds seen in a 15-mile-diameter circle (roughly the size of Georgia's smallest counties) are counted. The early counts were not as frequent as later ones and some were conducted on smaller areas. Although the surveys were few before the 1950s, the counts from 1959 to 1988 confirm the Breeding Bird Survey trends for several species that were counted in both (Table 6.2). The house sparrow, field sparrow, northern bobwhite (quail), northern flicker, hairy woodpecker, and eastern meadowlark decreased in both counts, whereas crows, blue-gray gnatcatchers, bluebirds, turkey vultures, great blue herons, the tufted titmouse, and the eastern phoebe increased. Many migrating birds and some residents appear in one of the surveys, but not the other. Of the sixty-two birds in Table 6.2, twenty-three increased and twenty decreased. For the other nineteen species the two surveys disagreed; they decreased in one survey and increased in the other, or did not change in one of the surveys. The results from these two long-running bird counts are mixed as one would expect, but there were slightly more gains than losses.

The eastern bluebird is a favorite bird of Georgia and of the United States. Several of our well-known songs exalt the "bluebird of happiness." The president of the Audubon Naturalist Society of the Central Atlantic States, Inc. says the bluebird "is American idealism personified—a flying piece of the sky, a living poem, a crystal note, an emblem of nature's moral conscience" and that "The bluebird ought to be America's national bird."[92] This bird adorned the cover of a book in 1928 entitled Some Helpful Georgia Birds, published by the Board of Game and Fish primarily for school children.[93] In spite of the love for the bird in the past and the attention lavished on it, many current Georgians probably would not recognize it.

The titles of two books about bluebirds in the last half of the twentieth century contain the word "survival," as if the bird was in danger,[94] and the Georgia Department of Natural Resources "initiated a Statewide bluebird restoration program" in 1988. An article on bluebird housing in a recent issue of the Farmers and Consumers Market Bulletin (2001) contained this statement: "Throughout the past half-century, loss of habitat contributed to the decline in bluebird populations." In fact, danger to the bluebird's survival has been a recurring topic throughout the twentieth century. In 1928, Hall and Rogers said "Formerly they were plentiful, but the freeze of 1898 destroyed great numbers of them." They were hopeful about their recovery, because they went on to say, "They are now increasing in numbers year by year." This statement thirty years after the freeze, indicates that this bird was not plentiful in Georgia in the first two decades of the twentieth century. According to Zeleny (1976), bird watchers in the 1930s and 1940s were asking, "Where are all those gentle bluebirds I knew 30 years ago?" Although this observation is not specifically about Georgia, such expressions as this and those by Hall and Rogers (1928) indicate that numbers of bluebirds have had noticeable ups and downs in the twentieth century. The reasons for these fluctuations are unknown, but probably complex.

A common assumption since the 1960s has been that the decline in bluebirds resulted from the use of insecticides such as DDT. However, as the foregoing observations indicate, large decreases in bluebirds occurred before DDT was used. Insecticides probably played a minor part, if any, in population changes. Zeleny (1976) gives six reasons for the decline in bluebirds, one of which was insecticides. He noted that cold winters in the South in which large numbers of bluebirds perished occurred in 1894–1895, 1939–1940, 1950–1951, and 1957–1958. Another cause for decline was said to be the loss of nesting sites. Woodpecker holes and other cavities in fence posts, dead tree trunks, and limbs are favorite nesting places. The decrease in small farms and the increased use of steel and chemically-treated fence posts has reduced nesting sites.

But Zeleny concluded that "in most areas undoubtedly the major cause" is the "overwhelming competition for nesting sites between bluebirds and the imported house sparrow (Passer domesticus) and starling (Sturnus vulgaris)." House sparrows (also called

English sparrow) were brought into the United States from England in 1851 and the European starling in 1890 and both have become numerous over the range of the eastern bluebird. The sparrow competes directly with the bluebird for nesting sites and actually evicts bluebirds from their nests anytime during nesting. The house sparrow is widely castigated in bird literature for its parasitism. The following poem gives a taste of the dislike.

> So dainty in plumage and hue,
>> A study in grey and in brown,
>
> How little, how little we knew
>> The pest he would prove to the town!
>
> From dawn till daylight grow dim,
>> Perpetual chatter and scold.
>
> No winter migration for him,
>> Not even afraid of the cold!
>
> Scarce a songbird he fails to molest,
>> Belligerent, meddlesome thing!
>
> Wherever he goes as a guest
>> He is sure to remain as a king.[95]

Likewise, starlings compete with bluebirds for nesting sites and according to Zeleny, starlings always win the competition. Davis and Roca (1995) were even more emphatic saying of the starling, "today it owns the natural tree cavities in the cities and farms." Both house sparrows and starlings were common in Georgia before mid-century.[96] House wrens also compete with bluebirds for nesting sites and some times destroy the eggs of bluebirds. The less competitive bluebird has been forced to leave that portion of its former habitat near human dwellings and to compete with these other birds in open fields where its chances are better. House sparrows, house wrens, and starlings like to stay close to human habitation. Early in the century the bluebird was apparently a common nester in backyards and neighborhoods in parts of the United States, but later restricted its nests to open spaces around fields and farmsteads.[97] However, Hall and Rogers (1928) indicate that in Georgia the bluebird may have never nested so close to houses and towns, saying they "will also nest in boxes when not too near human habitations."

Instead of a decline in bluebirds so often cited, the opposite is true for the last half of the twentieth century. The population increased by an average of 4.5 percent per year in Georgia from 1966 to 2000 in the Breeding Bird Surveys (Table 6.2). In Christmas Bird Counts it increased 1.9 percent per year. The numbers in Table 6.2 are average changes for the whole state, but the trend in population at Macon reinforces the increases in the last half of the twentieth century. In the first half of the century there was an average of seven bluebirds in Christmas Bird Counts at Macon. In the 1970s the average rose to fourteen, and in the 1990s to nintyone. Even in the Okefenokee National Wildlife Refuge bluebirds have increased from averages of four in three Christmas Bird Counts before 1950 to fify-two in the 1960s and eighty-nine in the 1990s. There have been efforts in recent years in some parts of Georgia to build nesting boxes for bluebirds and place them in open places along roadsides. This may have contributed to the increase in bluebirds, but it is not likely the main cause. The largest factor in the bluebird's recovery was likely a decrease in the bluebird's competitors and some improvement of habitat that we do not understand.

The house sparrow, its greatest competitor, decreased in both the Breeding Bird Survey and the Christmas Bird Counts (Table 6.2) and that of the starling did not change.[98] Robbins (1995) gave an interesting analysis of the decrease in the house sparrow: "Their numbers continued to expand until the automobile replaced the horse and the supply of waste grain was markedly reduced." The "waste grain" that the house sparrow consumed was bits and pieces that it scavenged from manure scattered along streets and roads. With its main competitor for nesting sites decreasing at a rate equal to or greater than the increase in bluebirds, it is understandable that the bluebird has made a comeback. What is not quite so understandable is the continued perception that the bluebird is declining.

There have recently been alarms raised about decreases in "neotropical migrant" birds that breed here and winter in the warmer climates of Central and South America.[99] In some parts of the southern Appalachians where the greatest declines were said to have occurred, neotropical migrants make up as much as 80 percent of the breeding birds. Concern about decline of these birds even reached the level of an official declaration of International Migratory Bird Day on 13 May 2000 and the enactment of the Neotropical Migratory Bird Conservation Act by Congress in July of that year.[100] President Clinton issued a message urging all Americans to celebrate the event and said, "Sadly, many species of these birds are suffering an alarming decline due to the degradation of their natural habitats. Polluted air, chemicals in fields, rivers and streams, toxic waste dumps, shrinking wild places—all have taken a heavy toll on their numbers."[101] At least some, if not most, of the concern is about wintering conditions in the Caribbean and Central and South America because the Neotropical Migratory Bird Conservation Act requires that 75 percent of future funds be spent outside the United States.[102] A more recent analysis than the one that caused such alarm indicates that as many neotropical migrant species have increased as have decreased.[103]

The trend is more positive for Georgia, with nine of thirty-six neotropical migrant birds increasing and only four decreasing since 1966 (Table 6.2). The remaining twenty-three did not change. Increasing and decreasing trends of different migrants are expected if they compete for the same habitat, and especially because all together they make up such a large proportion of the summer breeding bird population. It is unrealistic to expect that populations of all species would increase in a region already fully occupied by birds, or nearly so. It is also unrealistic to expect the population of every species to remain stable; more increases than decreases should be viewed as good news.

The species counted in the Breeding Bird Survey and listed in Table 6.2 as decreasing are apparently not in danger of disappearing any time soon, because to be included in the trend analyses they had to be encountered in at least fourteen routes. But species are sometimes claimed to be in danger when they are not, as is the case of the bluebird. Bachman's sparrow is listed as rare or uncommon and there has been concern expressed about its status.[104] The US Fish and Wildlife Service has considered Bachman's sparrow for the endangered species list and Gobris stated that populations of the sparrow have declined in its southern range. However, the North American Breeding Bird Survey does not indicate it as decreasing significantly in Georgia from 1966 to 2000. In fact, it was encountered in twice as many routes (thirty-one) as required for the analyses of the Breeding Bird Survey. The wood thrush is referred to as an "area-sensitive" species that has declined over much of its range over the past twenty years.[105] But the Breeding Bird Survey shows no change in Georgia from 1966 to 2000 during which time it was encountered in sixty-seven routes in the survey, the second highest number for any species.[106] Likewise the Breeding Bird Survey shows that nine of thirty-eight Georgia birds that Seabrook (2001) listed as being "in peril" actually increased since 1966, three significantly (the black-and-white warbler, the Kentucky warbler, and the yellow-throated vireo).

Because birds listed in Table 6.2 are rather plentiful, they are not the ones of most concern for loss or recovery. The more important trends are those of new species or endangered ones. Several birds have recovered from very low populations, but are not among those listed as increasing in Table 6.2. The wild turkey, bald eagle, peregrine falcon, and some species of duck have made notable gains, but were not common enough to be observed in sufficient early surveys. The ringed-neck duck was not listed in the Christmas Bird

Counts for Macon prior to 1976, but in the 1990s averaged 485 per circle, two and a half times as many as the cardinal. The cattle egret was listed as abundant in the Coastal Plain in the "Annotated Checklist of Georgia Birds" published in 1977[107] and has been stable in population since 1966.[108] It is a new species for Georgia, however, because it was not on a checklist published in 1945.[109] Many Georgians remember the first appearance in the 1950s or 1960s of these striking white birds following cattle around in South Georgia pastures, waiting for insects flushed from the grass.

The mix of cultivated fields, idle land, and farm forest that was prevalent in the 1800s and early 1900s might have been expected to increase the diversity of species. But for birds, the reverse appears to have been true for the rest of the twentieth century. Johnston and Odum (1956) described the increase in number and diversity of birds when fields were allowed to return to forest (Figure 6.7A). There were only 15 breeding pairs of 2 species (grasshopper sparrow and eastern meadowlark) per 100 acres in fields the first year out of cultivation. In old fields 25 to 30 years out of cultivation, pine forests were becoming established and there were over 100 breeding pairs per 100 acres representing over 10 species. In 100-year-old pine forest and 150-year-old oak-hickory woods the number of species more than doubled and breeding pairs were over 200 per 100 acres. This shows on a small scale what happened all across Georgia as the fields of the early 1900s reverted to forest. Although some have described the increasing pine plantations of Georgia as "barren,"[110] White et al (1996) found 70 to 75 percent as many wintering species in twenty- to thirty-year-old planted pines as in mature hardwoods or pines.

Along with the increase in bird species and populations with the shift from fields to forest, new species moved into the state. In addition to the movement of fourteen new breeding species into the Athens area described by Odum et al (1993), Oberle and Haney

(1997) found that seven species of northern forest birds moved into Northeast Georgia or extended their breeding range further south. In the opposite corner of the state, Crawford (1998) observed that in Thomas County seven species, including the cattle egret, have appeared since the mid-1900s because of range changes or introductions. Also seven species were recorded as breeding in the county that did not before, including the bald eagle. The balance of birds was described as follows: "On the debit side, we have lost only the

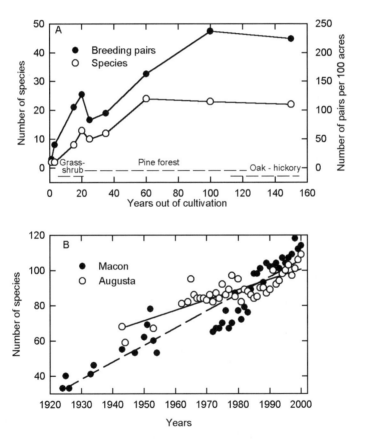

Figure 6.7. The change in the abundance of birds with increased time out of cultivation (A), and the increase in numbers of bird species in Christmas Bird Counts at Macon and Augusta during the twentieth century (B) Linear trend lines are shown in (B). Figure 6.7A redrawn with permission from Johnston and Odum (1956), data for B from Sauer et al. (1996) and the Georgia Ornithological Society (2000).

winter-resident Bewick's wren, but the rusty blackbird may be next. Nevertheless, we have gained much more than we have lost: Thomas County is richer ornithologically now than in the recent past." Thirteen species of birds were added to the Georgia "Regular Species List" in 1998 that were not on the "Annotated Checklist of Georgia Birds" in 1986.[111]

The movement of new birds into Georgia was not limited to these locations and instances, but seems to have been the case across the state. There was an average of 51.5 species encountered in the Breeding Bird Survey during the first five years, 1966 to 1970, and a significant increase to 57.7 species during 1996–2000. In Christmas Bird Counts begun in the first half of the twentieth century, there were upward trends in the number of species observed from the beginning. In Macon where counts began in 1925, there was an almost linear increase in the number of species throughout the rest of the century (Figure 6.7B). The increase for Augusta where counts began in the 1940s was similar, though not quite as steep. Although the two parts of Figure 6.7 are constructed quite differently, they may be portrayals of the same trends; the shift from small farms to a more forested landscape encouraged an increasing population and diversity of birds. In 1935, about 28 percent of land in the counties where Macon and Augusta are located was cropland and 40 to 45 percent was in forest. By the late 1980s forest occupied 65 or 70 percent of the land and harvested crops accounted for only about 3 percent.[112]

The building of watering places, ponds and lakes, all across the state has likely encouraged increases in many kinds of birds, but especially those that live on or near water. In A Birder's Guide to Georgia,[113] much of the focus for likely spots to see birds was on these watering places. Gulls, which used to be restricted to coastal areas are now common inland. The most common one, the ring-billed gull, was recorded only six times from 1925 to 1945—twice at Atlanta and Mount Berry and once at Athens and Augusta.[114]

At Macon, there were no gulls in the Audubon Society Christmas Bird Counts before the 1970s, but in the 1990s there was an average of 880 counted per year, representing three species, but mostly ring-billed gulls. At Nickajack Lake on the Tennessee border the average in the 1990s was 923 gulls per year. As mentioned earlier, the osprey that once was restricted to the coast has begun to nest near the large lakes in Georgia. There have certainly been drastic declines in some birds in Georgia during the twentieth century and a few, such as the passenger pigeon and ivory-billed woodpecker, have become extinct either early in the century or before. But in spite of frequent articles on the decline or endangerment of some species, it appears that Georgia's bird life has become substantially enriched and diversified, not poorer, in the latter half of the twentieth century.

Fire in the Forest

As discussed in chapter 2, native Americans burned forests of Georgia fairly often before the white man arrived. Of course, lightning caused some burning. The early settlers continued or intensified that practice. Early in the settlement of the state, fire was probably used to help clear the land for cultivation and to encourage the native grasses for grazing by cattle. In the twentieth century, burning of the forest has been one of the most environmentally controversial and persistent practices in Georgia. It has often pitted wildlife enthusiasts against foresters and neighbor against neighbor. Its popularity has fluctuated with drought conditions, land use emphasis, and wildlife management. Although the foremost authority on quail management[115] recommended burning of pinelands to encourage high populations of quail, Johnson (1948) advocated the opposite as the best for maintaining plentiful small game. In the 1920s the Georgia Department of Game and Fish (~1926) was adamant about the danger of burning forests emphasizing that "It is a universally recognized fact that *Forest Fires Kill more game annu-*

ally than all the hunters combined" [emphasis original]. After many years of indiscriminate burning that destroyed forests and prevented their regeneration, the campaign to prevent forest fires was successful (chapter 2), perhaps too successful. On the subject of burning to control undesirable vegetation, Dr. Eugene Odum concluded that "In view of the many questions now being raised about the use of toxic chemicals on a wholesale scale in nature it is important that controlled burning be considered as an alternate measure."[116] Rather that being the choice of the lesser of two evils, fire is usually chosen as the more economical.

The absence of fire in Georgia's forest has had consequences that were unforeseen and undesirable. Unburned forest accumulates large quantities of dead leaves and branches that in dry periods are fuel for wildfires more intense and destructive than those occurring every few years. During the devastating fires in the national forests of the western United States in 2000 and 2001, the policy of never burning to reduce combustible fuel was questioned. In addition, although "Smokey Bear" justified fire prevention largely to protect wildlife, it was soon realized that many species of wildlife depended on burning of the forest. The occurrence of fire in the southern pine forest for thousands of years naturally adapted species of birds and animals that not only tolerated burning, but required it. Burning of forests favors endangered species, the red-cockaded woodpecker, and those less in danger, such as the wild turkey and bobwhite quail.

The red-cockaded woodpecker was placed on the endangered species list in 1973 partly because burning of the forests decreased from the 1930s to late in the century. According to Jackson (1995), the decline of the woodpecker started with the wholesale cutting of the virgin pine and exclusion of fire from the old-growth pine forests further contributed to its decline. One problem for the red-cockaded woodpecker is the encroachment of hardwood trees that fill in the open spaces among the tall limbless trunks below the umbrella of pine foliage. These hardwoods shade out the herbaceous plants that cover the ground and supply food and refuge for the woodpecker. They may also encourage nesting of other birds that compete with the red-cockaded woodpecker. The US Fish and Wildlife Service has a recovery plan for this woodpecker that requires removal of hardwoods near nesting sites. Another uncommon species that has been considered for listing as endangered, Bachman's sparrow, requires the same habitat in pine forests.[117] Frequent burning of the woods favors pine trees and increases populations of both of these birds. King (1997) indicated that benefits of forest burning extend far beyond these two species: "mature pine stands in the southeastern U. S. managed by prescribed fire not only provide the critical habitat components sought by RCWs (red-cockaded woodpeckers) but also provide breeding habitat for a diverse avifauna community which may include sensitive species such as the Bachman's sparrow, common yellowthroat, prairie warbler, yellow-breasted chat, indigo bunting."

Abundance of quail and wild turkey, especially in South Georgia, has long been known to depend on burning of the woods. Open pineland with undergrowth of wild grasses and legumes is ideal habitat for quail.[118] Stoddard claimed that the quail "…was undoubtedly evolved in an environment that was always subject to occasional burning…." A mature stand of upland pine forest near Tallahassee Florida that had previously been burned regularly was not burned after 1966 and changes in wildlife and plants were recorded.[119] Forest cover increased from 43 to 91 percent in fifteen years and low-growing grasses and other plants decreased from 85 to 21 percent. Bird life on the unburned forest shifted from those of open habitat to woodland species, and total populations decreased by 44 percent. Brennan et al (1998) forecast increased prescribed burning on public land for forest and wildlife management, but further decreases on private forest land. They believe that the frequently burned public lands will serve as limited refuges for species, some of them endangered, that require open, park-like upland pine habitats.

Fish

Fishing is an even more popular sport in Georgia than hunting. According to the 1996 National Survey of Fishing, Hunting and Wildlife Associated Recreation[120] there were more than twice as many people who fish as those who hunt (1,088,000 compared to 403,000). Among women, fishing is far and away the more popular sport; 26 percent of the anglers in the state were women, but among hunters they were too few to list. The popularity of fishing is reinforced by the amount of money spent on boats, tackle, travel, and other expenses of the fishers. In 1996 the estimated spending was $1.1 billion in Georgia. This compared to $844 million spent for hunting. A little calculation reveals that the average hunter spent more money than the average fisher, but fishing is certainly a popular sport that helps Georgia's economy.

The freshwater fisheries of Georgia have been drastically altered by many factors. Three main ones are pollution (including sediment from soils), stream alterations (especially the building of dams), and introduction of non-native species of fish. A passage from Jenkins (1961) tells an interesting story that illustrates one of the changes.

> I might just mention in passing that one of the reasons proposed by the chartering committee that the University of Georgia be established at Athens in 1785 was that this site between the two branches of the Oconee River would provide excellent recreational possibilities for the students, particularly the annual spring shad runs. These would provide both recreation and food for the students. Of course, today a shad cannot make it up any of our rivers in Georgia over 20 miles. They are a fish of crystal clear waters and our streams no longer qualify in this regard. As a matter of fact, that is probably the understatement of the year.

The reference of Jenkins to shad was to the American shad, which were a favorite fish of early Georgia, and which migrated to the uppermost reaches of many of its streams to spawn. There are few large streams in Georgia that have not been dammed, hindering the migration of shad to many of their original spawning grounds.

Municipal and industrial dumping were the most obvious examples of ruining fish habitat by pollution. One of the poorest areas for fish at mid-century was the Chattahoochee River south of Atlanta. Gordon et al (1964) stated about this area that "Game fish have long been absent below the points of waste discharge." Mr. "Rock" Howard of the Georgia Water Quality Control Board described the Chattahoochee as "grossly polluted for about 100 miles below Atlanta from sewage and industrial wastes."[121] According to an Environmental Protection Agency (1980) report about the Chattahoochee in the late 1960s, "For at least 40 miles, the river was considered 'dead' because the high levels of oxygen robbing-wastes in the water made it difficult or impossible for fish to survive." Oxygen in that stretch of the river was consistently below the level required for most fish to survive (4.0 milligrams per liter) in 1963.[122] The lower Savannah River was in similar condition in the 1960s. For about a ten-mile stretch near the city of Savannah, dissolved oxygen during August 1967 through 1969 averaged 2–4 milligrams per liter.[123]

Many citizens believe fish kills to be of fairly recent occurrence and associated with pesticides and industrial dumping of chemicals. While industrial chemicals have accounted for large numbers of kills, many of them occurred long ago. Young et al (1950) tell of health problems in Macon that were unique to the early part of the century, but remark, "Yet others are quite modern in aspect as the problem of large quantities of dead fish floating in the river during July, 1902." Although the cause was never learned, public opinion attributed it to industrial dumping. From 1956 to 1960, the Georgia State Game and Fish Commission made thirty investigations of pollution incidents, most in response to fish kills.[124] At least twelve major fish kills were reported in 1963 "destroying thousands of pounds of game fish"; one below Lake Walter F. George on the Chattahoochee River extended 30 miles downstream.[125] Most of the fish kills and the worst instances of contamination were said to be due to industrial pollution and the Hastings and Frey report named thirty-seven industries and municipalities that were involved in cases they investigated from 1956 through 1960. The Rayonier pulp mill was said to have eliminated the fish in the Altamaha River for miles below Jesup when it began operations in the mid-1950s. The causes of the fish kills were corrected by 1962, but bacterial slime produced by the pulp wastes was so bad it nearly eliminated shad fishing below the pulp mill.

The good news is that these conditions have largely disappeared. The Chattahoochee River improved so much by the late 1970s that the Environmental Protection Agency (1980) used it as an example of progress. "Today the Chattahoochee is on the road to recovery, thanks to the efforts of the state, communities, industries, and EPA grants for municipal wastewater treatment facilities." The most telling assessment was a statement that Sidney Lanier "would once again be proud of his river." Perhaps the report writer for the Environmental Protection Agency got a little carried away, but his assessment was official. In the same report the Environmental Protection Agency reported that Sope Creek, a tributary to the Chattahoochee, was so polluted with sewage in 1973 that the sport fish population had been replaced with bloodworms, and water contact recreation had been banned. As a result of the clean-up, picknickers and fishermen returned along with bream, bluegills, largemouth bass, and trout. The assessment of the Chattahoochee River was confirmed eleven years later by Dozier (1991) who said "In fact, fish are now thriving in the Chattahoochee River south of metropolitan Atlanta in areas where they could not exist 15 years ago...." There are still fish kills in Georgia, but they must be much fewer and smaller than before, and they are much better investigated. The Georgia Department of Natural Resources (1999a) reported a total of thirty-eight incidents during 1996 and 1997, which killed 56,317 fish. Only four were found to be caused by industrial spills and six by sewage; each of these two causes accounting for 15 percent of the total. A little over half of the fish kills were of unknown cause. In seven of the incidents less than fifty fish were killed, and in three incidents less than ten were killed. Such small cases would hardly have been reported thirty years ago.

The return of fish to polluted waters in Georgia in the last half of the twentieth century is documented in numerous publications. As early as the 1970s the environmental clean-up was substantial. A report of the Georgia Department of Natural Resources (1976) contained the following: "It was learned in 1975 that fish were returning to areas of the Lower Savannah River where they had not been for many years. Successful fishing in the Conasauga River downstream from Dalton's wastewater discharge was also reported—a vast improvement over conditions five years ago." In a story about the clean-up of the Ochlockonee River in Southwest Georgia the *Associated Press* (1989) quoted John Carlton, a Moultrie lawyer who owns property along the river, "the redbreast bream are back, and small schools of striped bass have been seen this far upstream.... There's a sandy bottom once more, and all that algae and odor is gone." Although the Chattahoochee River south of Atlanta has been largely restored, fish populations there in the early 1990s were not quite like those before the river was polluted.[126] The proportion of game fish was less than in the Flint and Ocmulgee Rivers, but the bluegill, which was the fish most studied, appeared to be normal and healthy.

Although fish populations have increased where pollution was worst, pollution of some kinds and degrees improve the health of a fishery, at least as anglers see it. Much of the pollution of the past, although certainly not all of it, has a more positive name—fertilization. Recommendations for maintenance of farm fishponds have included fertilization since the 1940s. The same nutrients from Atlanta's waste, phosphorus and nitrogen, which were sent down the Chattahoochee River in large quantities to West Point Lake were added to farm ponds all across Georgia in the last half of the twentieth century. Speaking about West Point Lake, Jacobs (1998) said that in the early 1990s bass fishing suffered another decline, and "This time the culprit turned out to be clean water!" Jacobs was referring to the reduction of phosphorus released into the Chattahoochee River in the late 1980s.[127] An indirect statement of the desirability of some nutrient input above natural levels occurred in a report on the status of Blue Ridge Lake. "As a natural consequence of its relatively pristine watershed, Blue Ridge Reservoir has low concentrations of plant nutrients and therefore does not provide a food base adequate to support a well-balanced fish community."[128]

As remarkable as the accounts of fish recovery mentioned above are, they represent a small fraction of Georgia's fishery. Industrial and municipal pollution was never as bad in most streams as described for the lower Savannah and the Chattahoochee River south of Atlanta. Pollution, as bad as it was in some locations, may not have been the greatest cause of poor fish populations before mid-century. Many of the smaller streams of Georgia were strained of fish by "baskets" made of chicken wire or other materials, with wings that stretched across the channel and collected all of the fish that couldn't pass through the mesh. Seines were often used to drag many of the best fishing holes, collecting all except the smallest of fish. The poisoning of fishing holes with "green" black walnuts and even the use of dynamite were other illegal ways of collecting fish. Several articles in Outdoor Georgia in 1940 describe ways that fish in Georgia streams were decimated, including one instance of pollution of the Ochlockonee River by sewage from the city of Thomasville. One solution put forward by the magazine for the recovery of fishing was the building of artificial lakes and ponds. This, incidentally, was just at the beginning of the boom of lake and pond building in the state.

It is difficult to characterize the improvement of fish and fishing in Georgia because of the lack of documentation of fish populations in the past and lack of agreement as to what constitutes improvement. The "well-balanced fish community" described as unattainable for Blue Ridge Lake is a fishery manager's assessment; nature's balance is something else. Fishermen's tales about the abundance of fish are often as unreliable as those about the size of their catch. Wildlife experts also do not agree on the health of fisheries, and prevailing management practices have changed. An article in Georgia Game and Fish (1963) lamented the low population of game fish in the Oconee River, referring to a 1952 survey that showed only 2.77 percent were game fish, bass, bream, pike (pickerel), and crappie. "The balance of 97.23% was made up of carp, gar, suckers, and catfish.... The record revealed that 40% of the entire fish population of the Oconee consisted of roughnose suckers." The article did not say how extensive the survey was nor where on the Oconee it was located. The purpose of the article and an editorial in the same issue was apparently to solicit support for "Rough Fish Control Bill number 270" before the Georgia Senate to enable the Game and Fish Commission to control populations of undesirable fish in Georgia waters.

The inclusion of catfish in the "rough fish" category shows the inconsistency among fishermen and experts on the desirability of fish. The catfish is quite desirable to many fishermen. In 1985, almost as many fishermen in Georgia targeted catfish and bullheads as black bass (mostly largemouth bass), although by 1992 the number had dropped to 50 percent.[129] Frey (1981) classified catfish as "food fish" as opposed to the "game fish" designation for bass, crappie, bluegill, and sunfish. In fact, the catfish has really become a food fish. From an item of occasional meal for fishermen in the middle of the twentieth century, it has become a staple for supermarkets and restaurants. The commercial catfish that appears in supermarkets and restaurants is from catfish farms, though, not the natural waterways of Georgia.

The management of Georgia's fisheries has used both the removal of undesirable species and additions of desirable ones. In different cases the same species has been added and removed. There was an attempt to remove gizzard shad from Lake Blackshear in 1958 to increase populations of largemouth bass.[130] The population was so high they overwhelmed most of the other species and they were too large for their normal predators, such as largemouth bass. Rotenone, an early insecticide from plant sources, was sprayed on the lake and populations of gizzard shad were reduced from nearly 60 percent to less than 20 percent the next year. The percentage of bass in the lake and the catch rate by anglers increased for the next three or four years. The use of rotenone for this purpose apparently never caught on; the alternative of allowing bass to grow to large sizes to control the shad population was later used. In other cases gizzard shad and threadfin shad have been added to lakes as a forage fish for the same reason, to increase populations and size of bass. From 1953 to 1987, there were twenty-nine additions of bass, walleye (game fish), and threadfin shad to Lake Allatoona. The numbers of small fry added per year ranged from 2,500 in 1978 to a million or more in 1962, 1966, and 1967.

Black bass (largemouth, smallmouth, shoal, and spotted bass) are the most popular game fish in Georgia, being targeted by nearly twice as many fishermen as any other type. According to Jacobs (1998), bass fishing did not become a recreational pastime in Georgia until the twentieth century, but late in the century became "…a high-tech, fast-paced, and often competitive sport." The following whimsical description of bass anglers in a government fish survey report contains inferences and contrasts likely to incite more amusement and controversy than agreement.[131] "Bass anglers tend to be wealthier, more Southern and more male than other anglers." The description was surely meant to be serious, and whatever its accuracy, bass fishermen may have a bit more of the competitive "fever" than other fishermen. In 1999, there were over 800 bass fishing tournaments on Georgia's largest thirteen lakes sponsored by the Georgia BASS Chapter Federation, Inc. alone. There are certainly many more. In fact, bass fishing has truly become a pastime because many more are caught than are eaten. Quertermus (1998) shows that 90 percent or more of the hundreds of thousands of bass caught in tournaments each year are released alive. This is completely the reverse of fishing early in the twentieth century when nearly all fish were caught for eating.

Quertermus (1996–1999, 1998) has summarized results of bass tournaments in Georgia from 1978 to 1999 and Figure 6.8 shows the changes in fishing popularity and success in two of the lakes (Figure 6.8). It also shows the fluctuations of the fishery of West Point Lake caused by changes in management and possibly the cleaning of the Chattahoochee River. West Point Lake was built in 1975 and soon after bass fishing was "outstanding,"[132] as fishing usually is in new lakes. Before long bass fishing declined because the common prey fish, gizzard shad, had gotten so large small bass could not eat them, and the heavy population caused a decrease in production of small shad. In order to increase the size of bass in the lake so they could consume larger shad, the state Game and Fish Division raised the size of bass that could be kept from 12 to 16 inches in 1983. The larger size limit immediately affected bass fishing, reducing the number of tournaments from eighty-four in 1982 to twenty-nine in 1983, increasing the size of bass caught, and reducing the rate at which fishermen caught bass (at least the rate for those kept until weigh-in). The increased size limit was called a success, because a few years later the catch rate of bass increased again (Figure 6.8B). In contrast to the decreased catch in tournaments in the early to mid-1980s, Ager (1991) reported that bass

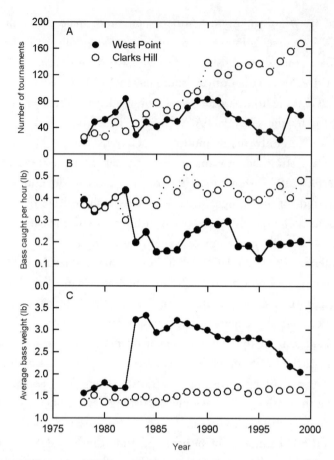

Figure 6.8. Changes since 1978 in the number of bass tournaments at West Point and Clarks Hill Lakes (A), weight of bass caught per angler (B), and average bass weight (C). From Quertermus (1998, 1996–1999). West Point tournaments that used 12-inch size limits for spotted bass are not included before 1998.

harvest by other anglers on West Point Lake reached an all-time high in 1985. But in the early 1990s, the success rate decreased again; this decrease is the one Jacobs (1998) blamed on "clean water."

It is hard to know whether the number of tournaments at West Point was related to the success of bass fishing. The number of tournaments was slightly higher on West Point than Clarks Hill Lake prior to 1983, but after the size limit was increased for bass it was always less. Although the number increased again to about eighty in the late 1980s, it has declined after that. The decline in number of tournaments on West Point Lake was not quite as drastic as shown in Figure 6.8, because only those tournaments in which a 16-inch limit was used for largemouth and spotted bass are included from 1983 to 1997. As spotted bass increased in abundance in the 1990s, more clubs started using a 12-inch limit for this fish. When those tournaments were included, the totals were sixty-seven for 1998 and fifty-nine for 1999 (Figure 6.8A). In contrast to West Point, tournaments at Clarks Hill Lake have increased steadily from 25 in 1978 to 160 in 1999. The pounds of bass each angler caught per hour on West Point Lake since the 16-inch size limit was imposed has been half that on Clarks Hill Lake, but the average size of bass caught has been much larger (Figure 6.8B and 6.8C). The average size of the largest bass caught in tournaments on the two lakes has been almost the same in the 1990s. So popularity of the two lakes for tournaments is apparently based on something other than the size of the fish.

Although bass tournaments are not completely reflective of the fishing in Georgia they do show that fishing in the state has improved tremendously in the last half of the century. The number of bass tournaments has increased on all of Georgia's large lakes, except West Point, in the last thirty years. The days of the bamboo pole, a can of worms on a creek bank, and fried perch (sunfish) or catfish for supper bring back happy memories, but they were no more enjoyable for fishing than today and were much less productive of edible fish. Nobody free of the veil of nostalgia, would trade today's fishing for "the good old days" of the first half of the twentieth century.

Seafood

Seafood is an important food category for the coastal industry of Georgia, for the restaurant trade, and for home consumption. There are occasional reports of the depletion of coastal seafood resources due to over-fishing, pollution, or weather. The available records indicate drastic over-harvest of several species early in the century. The trend in total commercial seafood harvest shows a rapid rise from 1890 to the late 1920s and a drastic drop in the early 1930s (Figure 6.9A). There do appear cycles of increase and decrease of total harvest over the years, but no long-term trend since the 1930s.

Over-fishing of the estuaries and coastal rivers has been a long-standing problem. In the 1930s a report by the East Georgia Planning Council (1937) revealed that from 1929 to 1934 the catch of shrimp decreased from 12.4 to 6.8 million pounds, while the number of shrimp boats of 5 net tons or more doubled from twen-

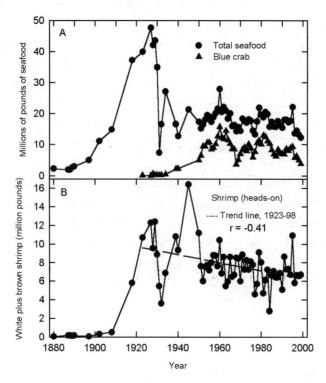

Figure 6.9. Changes in commercial seafood harvests in Georgia since 1880. Total seafood and blue crab (A) and shrimp (B). From the East Georgia Planning Council (1937), Carley (1968), US Department of Commerce (1973–1980), and the National Marine Fisheries Service (2001).

ty-two to forty-four. This decrease in shrimp harvest is shown in (Figure 6.9B), which also shows that the catch more than doubled again by 1945. The overall trend in shrimp harvest since 1923, just before it reached its first peak, is an erratic but significant decrease (Figure 6.9B); since the 1950s, it is hard to say whether shrimp harvest has changed. The Georgia Department of Natural Resources (1999a) reported that "White shrimp landings have varied over the last 40 years with no overall trend." The forty-year trend in brown shrimp was said to have been downward, but brown shrimp make up less than 20 percent of the total.

The report of the East Georgia Planning Council also revealed that oyster production decreased by 93 percent from 1908 to 1934 (Figure 6.10). Even that decreased oyster catch was still important in the 1930s as shown by the following statement about this and the total seafood industry: "By comparison with former years, the oyster industry of Georgia has almost ceased to exist; yet the value of this almost extinct industry amounts to 11% of the total fisheries of Georgia, which proves the present deplorable state of this entire industry."[133] From that peak of over 6 million pounds in 1908 oyster harvest has decreased to only a few thousand pounds in the 1980s and 1990s (Figure 6.10), but most of the decrease occurred before the 1920s. Reasons for the drastic decrease in harvest of oysters early in the century were several, including economic ones, but over-harvesting without any replenishing of the beds was likely a main one. The East Georgia Planning Council quoted from a letter of the US Bureau of Fisheries that "The deplorable depletion of oyster bottoms in Georgia waters is entirely due to over-fishing and the lack of oyster cultivation." Over-fishing was also indicated by the following from the Georgia Department of Natural Resources (1938), "A start has been made toward establishing new oyster beds and in restoring old ones to productivity."

Over-fishing of shad, probably the most prized fish of early Georgia, decimated the supply early in the century. The depletion

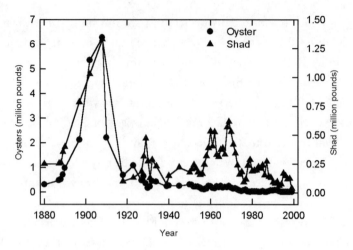

Figure 6.10. Changes in oyster and shad harvests in Georgia since 1880. From the East Georgia Planning Council (1937) Carley (1968), US Department of Commerce (1973–1980), and the National Marine Fisheries Service (2001).

was described in the following way. "Considering the high price this fish brings, which should intensify effort towards its production, the obvious conclusion is that the supply is being depleted at an alarming rate."[134] However, the supply was apparently depleted two decades before this report. After reaching a peak of 1.33 million pounds in 1908, the catch declined 90 percent, to only 100,000 pounds in 1918 (Figure 6.10). For the rest of the century commercial landings averaged only about 250,000 pounds per year. Although the apparent conclusion is that over-fishing caused the early decline in Georgia, many other factors including labor and economic problems and governmental restrictions on harvest may have kept the landings low in the second half of the twentieth century.

There was serious pollution of some estuaries that resulted in restrictions of fishing and harvest of oysters in the 1960s and 1970s, but actually serious pollution, and restrictions on harvest, started much earlier. The report of the East Georgia Planning Council (1937) describes some of the pollution early in the century. "While sewage pollution is a serious potential problem, there is also the menace of industrial waste and other chemical pollution that seriously affects propagation, growth, and supply of all species of fish life." The report went on to say "With the ever increasing industrial development in this area, it is necessary and advisable that a study be made of the problem of industrial waste pollution, and a constant check maintained on this potential menace."

The sewage pollution was serious enough in the 1930s that oyster beds in much of the estuaries around Savannah and Brunswick were declared off limits based on "bacteriological analysis." The report of the East Georgia Planning Council contained a passage about lack of restoration of the oyster beds that today seems odd for its economic justification.

> The polluted areas along the Georgia coast comprise only a small fraction of the total actual or potential oyster growing areas. The elimination of pollution in the condemned areas, arising from discharge of sewage from cities located on tributary rivers would cost many millions of dollars. It is questionable, therefore, whether elimination of existing pollution would be economically justified by the degree of benefit resulting to fisheries, as well as to recreation. However, the problem does offer a long time potential threat, as the doubling or tripling of the existing pollution load might result in serious consequences.

As far as the records show, the extent of polluted waters did not justify cleanup, because it was about thirty years later that the reversal of sewage pollution began. By that time, both the sewage and industrial pollution had gotten much worse. A report on the management of the Georgia Game and Fish Department by Gordon et al (1964) stated that pollution was a major problem facing the shellfish industry and that seventy-four locations and about 46,900 acres of estuary were closed to shellfishing because of pollution. These were mostly the same areas closed for harvest in 1934. Addition of another 4,300 acres was recommended in 1970,[135] although it is not clear that the prohibited acreage had remained stable through the 1960s. As bad as the pollution was then and earlier, the steep decreases in harvest of shad and oysters from 1908 to 1920 and of total seafood a decade later, probably resulted from depletion of the supply rather than pollution, which would likely have had a much more gradual effect.

Trends in the shellfish producing areas of Georgia and the extent of pollution are clouded by the changing statistics describing the estuaries and changes in the regulations and reporting. The Georgia Department of Natural Resources gave the "potentially productive shellfish growing area" as 496,000 acres in 1995, but 700,000 acres "of potential shellfish habitat" in 1997.[136] These figures account for essentially all of the Georgia marsh, but the shellfish growing areas are much less. Ten percent of the potential 700,000 acres was described as producing viable shellfish stocks in 1997,[137] but even this 70,000 acres appears to include areas other than the natural beds where oysters are most plentiful. Linton (196?) reported that a study in 1966 and 1967 showed that there were only "10,182 acres of intertidal oyster beds" where most of the oysters occurred.

During the 1980s and 1990s classification of the shellfish areas for pollution seem to have been in a state of flux. In 1995, 35 percent of the potential area, about 174,000 acres, was listed as "approved," 92,000 acres as "restricted," and 232,000 acres as "prohibited." In 1997, only two classifications were used ("approved" and "prohibited") and only 31,816 acres were listed as "approved." Adding to the confusion, the *Associated Press* reported that only 14,133 acres of the Georgia coast are "legal" for shellfish harvest.[138] The uncontaminated acreage did not likely decrease so drastically in so few years, but the "approved" acreage decreased because the regulations changed.

The extreme precautions against health problems of contaminated shellfish have produced three restrictions unheard of in earlier times.[139] First, 30,000 of the 92,000 restricted acres in 1995 "have the classification because of bacteria levels which occur naturally as a result of local wildlife." Second, 77 percent of the 232,000 "prohibited" acres were so classified because they had not been tested enough to exclude them. Third, areas such as those around marinas are classified as prohibited because they have the potential for contamination, whether or not they actually are. The acreage closed for shell fishing appears to have drastically increased since the 1960s, from 46,900 acres[140] to 232,000 acres in the early 1990s.[141] Because of increasingly stringent standards, the acreage off limits to shellfish harvest in the 1990s was several times higher than in the 1960s when water pollution was much worse.[142]

Realities and Perceptions

Wildlife used to be considered largely a byproduct or an adjunct of agriculture. The birds and animals that lived in the fields or nearby woods were hunted if they were needed for food or if they interfered with farming. They were also sometimes hunted for sport or mischief. Young boys of farm and town would often shoot any wild animals or birds that appeared with sling-shots, BB-guns, or rifles. There were frequent admonitions against shooting songbirds,[143] but they were often the object of the hunt. It is hard to believe that such juvenile hunting actually had much effect on wildlife populations. Whatever the effect, wildlife was considered by most a part of country life and hunting was one of the more enjoyable pastimes of the early twentieth century.

The perception at the end of the twentieth century is quite different. Many consider wildlife as an indicator of the health of the environment, to be protected rather than exploited. Contrast the concern for eagles today with the fact that more than 100,000 eagles were killed in the Alaskan Territory between 1917 and 1950 for bounties ranging from $.50 to $2.00.[144] The possible extinction of an exotic species causes as much alarm today as bobcats or chicken hawks preying on farm animals did sixty or seventy years ago. The concern and support for wildlife preservation is shown in many ways at the end of the twentieth century. There are today many vocal support groups for individual species as well as for wildlife in general. These groups not only press for legislation and funding for their favorite species, but raise money to support research and education. Citizens not belonging to these groups also support wildlife promotion by buying car license plates or checking off dollars for wildlife on their income tax returns.

The funds from license plates and tax check-offs are used to support research and protection for endangered and other nongame species. Support for such protection is widespread among the population, but according to Godbee (1993) when the financial burden shifts from the public to the individual attitudes change. He faulted the Endangered Species Act for relying on confrontation rather than cooperation. The conflict works this way—if landowners encourage an endangered species to occupy the land, their property becomes an endangered species preserve, and diminished in income producing potential. This disincentive apparently causes some landowners to manage in such a way as to discourage endangered species. Godbee calls such action both undesirable and understandable, and suggested that the protection of endangered species would be helped by incentives for landowners to "grow" habitat. Such an approach is now being used in Georgia to encourage increases in quail.

An important pastime today is the watching and feeding of wildlife. In 1985, the US Fish and Wildlife Service estimated that 2.7 million or about 50 percent of Georgia's population participated in "nonconsumptive activities" related to wildlife.[145] These included observing, photographing, and feeding birds and animals. Most of these nonconsumptive activities consist of watching and feeding at home, but in 1996, 639,000 Georgians traveled away from home to watch, photograph, or feed wildlife.[146] It is not clear whether this more passive interest in wildlife has decreased in the last decade or so. Although surveys from 1985 and 1996 may not be comparable (techniques changed between the surveys) they show that 49 percent of Georgia residents fed birds in 1985 and 26 percent in 1996.

Wildlife in the twentieth century is a story of changes in landscape, populations, and attitudes. Trends for wildlife in Georgia are so complicated and so interdependent that a conclusion about improvement depends on one's biases. There is no question that wildlife in Georgia is more diverse and the large game animals more plentiful than they have been in more than a century. The question is whether trading quail for deer and rabbits for bobcat is improvement. The trade does not have to be complete and almost never is. That is, the trading of the small farm habitat ideal for quail, rabbits, and some songbirds for large mechanized farms and extensive forests has been complete enough to diminish but not endanger these species. The restocking of those prized game species of early Georgia is one of the great environmental successes of the twentieth century. Songbirds have not decreased during the century, on the contrary, they have become more diverse. The beloved bluebird has increased significantly since the middle of the twentieth century. The bobwhite quail has decreased since the days of small farms and "King Cotton," but has been replaced in visibility and interest by deer, wild turkey, and waterfowl that were once so scarce in Georgia. The mourning dove has become an urban bird, decreasing during the breeding season, but increasing in the winter over the last half century.

Rescue of diminished or endangered species is not without cost to other species, and sometimes the cost is born by other endangered species. Conflicts between alligators, snail kites, and wood storks illustrate the complications of saving endangered species, and the effects of changing habitats on the balance of species. When alligators were restocked to increase their numbers, their increased predation on apple snails, the preferred food of the kite, further endangered the snail kite. The snail kite requires a habitat with long periods of inundation to ensure adequate populations of the apple snail. The wood stork, however, requires extensive dry-down periods for nesting.[147] So three threatened species compete in the web of life, and efforts to help one hinders the other.

Most Georgians believe that the National Forests and National Wildlife Refuges are teeming with wildlife and that the neighborhoods are relatively sterile. For some game and other large animals it is true, but birds are more numerous and diverse in urban areas, as is shown by the Christmas Bird Count of the National Audubon Society.[148] Atlanta had more Canada geese, bluebirds, and crows than any other of the seventeen sites at which they were counted from 1991 to 1999 (Table 6.3). Atlanta also had more pileated woodpeckers, those large crow-sized woodpeckers so exciting to spot in a neighborhood, than the Chattahoochee National Forest. None of the National Wildlife Refuges, except the one on St. Catherine's Island on the coast, had appreciably more than Atlanta. Some birds are, of course, more numerous in forests and wildlife reserves, but the number of species in recent Christmas Bird Counts was greater near all of Georgia's towns, except Dalton, than the Chattahoochee National Forest and the Okefenokee National Wildlife Refuge. The locations along the coast had the largest numbers of species, probably because of the mixture of sea, marsh, and land birds. Perceptions of towns and cities as unfriendly places for birdlife, either because of pollution or degraded landscapes, are not entirely correct. The negative perception is likely correct for inner-city areas, but certainly not for suburbs.

If wildlife is more numerous and diverse than it used to be, have we increased populations of the most desirable species? The answer is it depends on the interests of the individual considering the question. Some people may have equal interest in all species, but it is unlikely. The demise of a popular bird like the bobwhite quail, the eastern bluebird, or the bald eagle would be considered a catastrophe by many, but few would miss the crow, the armadillo, or the rattlesnake. One factor that determines whether an animal is considered beneficial is its utility or trouble. Farmers with crops and livestock to protect do not value the same wildlife that suburbanites do, although the more wild animals move into the suburbs, the clos-

Table 6.3. Number of birds in a 15-mile-diameter circle in 17 Georgia locations. From the Christmas Bird Count (Georgia Ornithological Society, 2000). Averages for years 1991-1999.

	Canada goose	American crow	Mourning dove	Bluebird	Pileated woodpecker	Number species
Dalton	102	116	97	41	3	65
Athens	193	536	376	243	12	84
Atlanta	463	857	448	355	24	85
Peachtree City	56	503	232	280	17	85
Augusta	45	79	366	24	11	94
Callaway Gardens	164	447	207	154	19	80
Macon	126	200	422	91	19	101
Columbus	71	234	283	48	9	92
Albany	111	179	227	69	10	87
Bainbridge–Lake Seminole	25	155	92	23	4	74
Chattahoochee NF[1]	25	226	37	24	8	58
Piedmont NWR	302	285	116	166	22	101
Okefenokee NWR	13	72	43	89	22	71
Sapelo Island	0	87	98	88	26	120
St. Catherine's Is.	0	81	99	95	42	129
Harris Neck NWR	0	134	109	50	20	120
Glynn County	24	91	563	22	14	139

[1] NF = National Forest; NWR = National Wildlife Reserve.

er their attitudes become. The whole population, as represented by the state, has its favorite species. One can tell by the animals and birds on which the most money is spent.

Emphasis by state wildlife officials has changed over time, as can be seen by the change in name from the "Department of Game and Fish" early in the century to "Wildlife Resources Division" at the end. When game was nearly the only interest of the populace, the state emphasized protection and regeneration of those species. When the interest changed from hunting game to protection of all creatures and habitat, emphasis by the state changed. Contrast recent programs to research and protect nongame species to a passage in a publication of the Department of Game and Fish in the late 1920s. "Practically all of the birds in the state are protected by game laws. However, there are a few birds that are not protected. The hawk, eagle, owl, crow, sparrow, and meadow-lark are considered to do more harm than good and may be shot at any time."[149] This sounds like an invitation to get rid of these species, long before the modern reasons for species decline—pollution, pesticides, and habitat destruction. To science, conservation of all animal species is important for the diversity of nature and the preservation of gene pools that some day may be useful to man. But for mobilization of popular opinion, easily recognized and loved species are much more valuable than irritating and ugly ones.

For many, attitudes about wildlife are shaped by mediatypes. Story books, cartoons, movies, and entertainment television have "lionized" the common, the exotic, and the dangerous almost at random. Bears are made to look friendly, lovable, and protective of the forest. In reality, wild bears are none of these. Even alligators can be made to look child-friendly, but their instincts haven't changed. Sooner or later, the dangerous or inconvenient nature of wild animals is impressed on the adult. The raccoon of storybooks is either cuddly or mischievous and even the opossum of Pogo-land is lovable, but those that scatter the garbage regularly are a nuisance. It is easier to love the image than the reality. Farm children, or those that spend a lot of time in the field and forest, learn early to separate the lovable from the dangerous.

Declines and endangering of birds are frequently blamed on DDT or other pesticides. Articles on both the bald eagle and peregrine falcon cite DDT as a major factor in their decline. A news release of the Georgia Department of Natural Resources (2000a) claims that, "Once fairly common, bald eagles suffered significant declines in the 1950s and 1960s primarily as a result of Dichloro Diphenyl Trichloroethane (DDT) poisoning." However, both the eagles and falcons were listed as rare in Georgia in records going back to the turn of the century, before such pesticides were used. Although Greene et al (1945) and Burleigh (1958) described the bald eagle as fairly common along the coast before the 1940s, Denton et al (1977) said it was "Formerly a rare resident along the coast." It was rare or uncommon in the rest of the state. Two prominent bird watchers, Herbert L. Stoddard, who studied birds in Grady County in Southwest Georgia,[150] and Milton N. Hopkins in Ben Hill County in South Central Georgia[151] reported only five sightings between them from the 1920s to the 1970s. In a biological reconnaissance of the Okefenokee Swamp from May to July 1912, Wright and Harper (1913) did not see bald eagles or their nests, but were told their appearance was "occasional." They noted that "It apparently does not breed here." Burleigh (1958) noted that "it has been seen from time to time in the interior of the state where it is not known to nest."

These observations indicate scarcity of bald eagles in Georgia in the first half of the century, but actual counts indicate the same. In 86 Christmas Bird Counts conducted at various locations and years before 1945 there were only 5 bald eagles observed—a rate of 5.8 eagles per 100 bird counts. During the period when DDT was most widely used, 1945 to 1975, there were 17 observed in 264 counts; a rate of 6.4 eagles per 100 counts. So, it appears that bald eagles were no more plentiful in Georgia prior to the use of DDT than during the time it was widely used. Since 1975, about the time DDT was banned and eagle restoration efforts began, the rate has been 77 eagles seen per 100 Christmas Bird Counts. The numbers in the 1990s were much higher; 123 per 100 counts. Although bald eagles are much more plentiful in Georgia now than in the past 100 years, they were rare before the use of DDT and other similar pesticides.

The peregrine falcon was more rare. A single nesting was recorded for Georgia in Dade County in April 1942. According to Berger et al (1969), for both Georgia and Alabama, "Another has not been seen to be active since before the turn of the century." A single nesting of the falcon and so few sightings of the bald eagle in the first half of the twentieth century make it unlikely that the use of DDT, unavailable until after 1945, caused the decline of these birds in Georgia. DDT and other pesticides certainly killed birds and DDT was found to weaken eggs of several species and caused some bird declines in that way. It is quite another thing to extrapolate to the decline of populations of many bird species.

Apart from the concern about endangered species, the hallmarks of successful wildlife management are abundance and diversity. But the sufficiency of both depends on perception. Recovery of some species is hailed, but the appearance or increase of others is deplored. The more bluebirds the better, but bird enthusiasts would like to see fewer of the brown-headed cowbird, the house sparrow, and the starling, because they compete with better-loved birds. Diversity has been increased by movement of several species into Georgia in the last half of the twentieth century. All newcomers have not been welcomed but most have been accepted, even though there must be some adverse consequences for wildlife of similar habitat. From the south have come the cattle egret and armadillo. Although their effects may not have been as drastic as that of the boll weevil and the fire ant, which came from the same direction, they certainly do have some effect. Coyotes have entered every county in the state. They compete with fox for similar habitat and food. High populations of coyote can't be helpful to the fox nor rabbits. Rabbits are a favorite food of both.

So, although the accomplishments in wildlife restoration of the twentieth century do not suit everyone, they are impressive. Outstanding recoveries of large game species have occurred with little, or only moderate, depletion of others. Decreases of songbirds of open fields and fencerows have been offset by increases in forest and suburban birds. Diverse species of native fish now inhabit streams that forty years ago were so polluted none, or only "rough" fish, lived there. Tens of thousands of lakes and ponds that did not exist sixty years ago now have abundant fish, birds, and other aquatic animals. Hopkins (1958) concluded after construction of about 30,000 ponds that "These farm ponds have and in the future will bring about marked changes in birds, habits and habitats and will add importance to the Atlantic waterfowl flyway." Indeed they have. In 1996, 82 percent of the days spent fishing in Georgia were on lakes and ponds.[152], up from 42 percent in 1975.[153] Release of cold water from the bottom of some of Georgia's largest lakes has created trout streams where water used to be too warm for trout. Some of the larger lakes now attract birds that formerly were restricted to immense wetlands along the coast. Endangered species, including the alligator, wood stork, bald eagle, and the peregrine falcon, have made remarkable comebacks. The hopeful assessment of Jenkins at mid-twentieth-century (1953) that "Georgia could become one of the very best states in the United States in regard to wildlife production" is probably closer to reality than he or others at the time dared hope.

[1] Stanwyn Shelter in Zeleny, 1976.
[2] Jenkins, 1953.
[3] Rostlund, 1960.
[4] Wetherington, 1994.
[5] Whitney, 1994.
[6] Johnson, 1948.
[7] Jenkins, 1953.
[8] See chapter 2.
[9] Allen, 1948.
[10] Jenkins, 1952.
[11] Georgia Department of Natural Resources, 1938.
[12] Jenkins, 1953.
[13] US Department of the Interior and US Department of Commerce, 1998.
[14] Whitney, 1994.
[15] Montgomery, 2001.
[16] Williams, 1999.
[17] Associated Press, 1999.
[18] Georgia Department of Natural Resources, 1998b.
[19] Jenkins, 1953.
[20] Carter, 1994.
[21] Thackston et al, 1991.
[22] Georgia Department of Natural Resources, 2000b.
[23] Landers and Mueller, 1986.
[24] Jenkins, 1953.
[25] Brunswig and Johnson, 1973.
[26] US Department of the Interior, 1989.
[27] US Department of the Interior and US Department of Commerce, 1998.
[28] Sauer et al, 2001.
[29] Sauer et al, 1996.
[30] Hardie, 1999.
[31] Albany Journal, 1940.
[32] Dolton, 1995.
[33] Baskett and Sayre, 1993.
[34] Dolton and Smith, 1998.
[35] Sauer et al, 2001.
[36] Sauer et al, 1996.
[37] Jenkins, 1953.
[38] Daniel, 1949.
[39] Georgia Department of Natural Resources, 1998c.
[40] Jenkins, 1953.
[41] US Department of the Interior and US Department of Commerce, 1998.
[42] US Department of the Interior, 1989; US Department of the Interior and US Department of Commerce, 1998.
[43] US Department of the Interior and US Department of Commerce, 1998.
[44] Council on Environmental Quality, 1997.
[45] Hicks, 1977.
[46] Jenkins, 1953.
[47] Wharton, 1977.
[48] Pullen, 1971.
[49] Lewis, 2001.
[50] Hon, 1979.
[51] Georgia Department of Natural Resources, 1993.

[52] Georgia Department of Natural Resources, 1993.
[53] Laerm, 1994.
[54] Georgia Department of Natural Resources, 1993.
[55] Georgia Department of Natural Resources, 1938.
[56] Georgia Department of Natural Resources, 2001b.
[57] Scott, 2000.
[58] Georgia Department of Natural Resources, 1993.
[59] Associated Press, 2001a.
[60] Georgia Department of Natural Resources, 1993.
[61] Ruckel, 1987.
[62] Georgia Department of Natural Resources, 1993.
[63] Georgia Department of Natural Resources, 1993.
[64] Johnson and Kennamer, 1975.
[65] Georgia Department of Natural Resources, 1993.
[66] Georgia Department of Natural Resources, 1993, 1999b.
[67] Georgia Department of Natural Resources, 1993.
[68] US Department of the Interior, 2000a.
[69] Buchan et al, 1998.
[70] Plummer, 1997. See Chapter 4.
[71] US Department of the Interior, 1989.
[72] Martin et al, 1991–1994.
[73] Georgia Department of Natural Resources, 1999c.
[74] Georgia Department of Natural Resources, 1998d.
[75] Council on Environmental Quality, 1994.
[76] Knight Ridder Newspapers, 1999.
[77] Georgia Department of Natural Resources, 2000a.
[78] Seabrook, 1998a, b.
[79] McCort and Coulter, 1991.
[80] Harris, 1993.
[81] Sauer et al, 2001.
[82] Blackshaw and Hitt, 1992.
[83] Bailey, 1997.
[84] Hill et al, 1987.
[85] Peterson, 1979.
[86] US Department of the Interior and US Department of Commerce, 1998.
[87] Whitney, 1994.
[88] Odum et al, 1993.
[89] Sauer et al, 2001.
[90] Sauer et al, 2001.
[91] Hopkins and Odum, 1953.
[92] Zeleny, 1976.
[93] Hall and Rogers, 1928.
[94] Zeleny, 1976; Davis and Roca, 1995.
[95] Mary I. Forsyth in Green, 1933.
[96] Greene et al, 1945.
[97] Davis and Roca, 1995; Zeleney, 1976.
[98] Sauer et al, 2001.
[99] Simons et al, 1999; Peterjohn et al, 1995.
[100] US Department of the Interior, 2000b,c.
[101] US Department of the Interior, 2000c.
[102] US Department of the Interior, 2000b.
[103] Peterjohn et al 1995.

[104] Gobris, 1992.
[105] Simons et al, 1999; Graham, 1990.
[106] Sauer et al, 2001.
[107] Denton et al, 1977.
[108] Sauer et al, 2001.
[109] Greene et al, 1945.
[110] Hopkins, 1958.
[111] Anon, 1998.
[112] US Department of Agriculture, 1946; Georgia State Planning Board, 1939; Bachtel, 1990; Georgia Department of Natural Resources, 1996a.
[113] Blackshaw and Hitt, 1992.
[114] Greene et al, 1945.
[115] Stoddard, 1931.
[116] Odum, 1963.
[117] Gobris, 1992.
[118] Stoddard, 1931.
[119] Brennan et al, 1998.
[120] US Department of the Interior and US Department of Commerce, 1998.
[121] US Department of the Interior, 1966.
[122] US Department of the Interior, 1966.
[123] US Department of the Interior, 1969.
[124] Hastings and Frey, 1962.
[125] Gordon et al, 1964.
[126] Mauldin and McCollum, 1992.
[127] See chapter 4.
[128] Tennessee Valley Authority, 1990.
[129] U. S. Department of the Interior, 1989; US Department of the Interior and US Department of Commerce, 1998
[130] Zeller and Wyatt, 1967.
[131] Pullis and Laughland, 1999.
[132] Jacobs, 1998.
[133] East Georgia Planning Council, 1937.
[134] East Georgia Planning Council, 1937.
[135] Olmsted et al, 1970.
[136] Georgia Department of Natural Resources, 1997, 1999a.
[137] Georgia Department of Natural Resources, 1999a.
[138] Associated Press, 2001b.
[139] Georgia Department of Natural Resources, 1997.
[140] Gordon et al, 1964.
[141] Georgia Department of Natural Resources, 1997, 1999a.
[142] See Chapter 4.
[143] Hall and Rogers, 1928.
[144] Barnes, 1951.
[145] US Department of the Interior, 1989.
[146] US Department of the Interior and US Department of Commerce, 1998.
[147] Harris and Frederick, 1991.
[148] Georgia Ornithological Society, 2000.
[149] Georgia Department of Game and Fish, -1926.
[150] Stoddard, 1975.
[151] Hopkins, 1975.
[152] US Department of the Interior and US Department of Commerce, 1998.
[153] Geihsler and Holder, 1980.

CHAPTER SEVEN

Summing Up

Georgia's environment, judged by nearly all objective measures, is better at the end than early in the twentieth century. Streams are cleaner, forests are more abundant and productive, the air is clearer, and wildlife is more plentiful. The result is that we are healthier, our countryside and suburbs are more beautiful, and our outlook should be much brighter because of it. Without careful study it may be the conclusion of some that Georgia's best days lie somewhere in the past. In regard to the environment, it is clearly not true for the twentieth century. In an essay entitled "The New South" about the benefits of small, diversified farms, Sidney Lanier wrote, "small minds love to bring large news, and, failing a load, will make one."[1] Environmental progress in Georgia is "large news," but the "load" I have brought is not artificial. I hope the previous chapters have conveyed the reality that our environment has become more livable for us and other creatures in the last half of the twentieth century.

Despite the virtues that Sidney Lanier and others claimed for small farms, the subsistence farming from the Civil War to the Great Depression was no friend of the environment. Devastation of the land that Lanier and Henry Rootes Jackson wrote about in the 1800s is not a thing of the past, but a legacy for generations to come. The good news about the land is that the healing began seventy years ago and we deal now, and hopefully in the future, only with the scars, not the open wounds. The scars are the eroded hillsides and gullies that have been healed and the silted-in streambeds that will continue to deliver silt to reservoirs for many years. In the middle of the great depression, Luther F. Elrod, wrote the following prophecy in the Sandy Creek News. "It is not too visionary to say that, if the farmers of Georgia will follow a wise land use program, in another generation the hills will again be covered with valuable forests; meadow lands will pasture fat cattle; streams will run clear and be filled with fish; woods will be filled with abundant game; and painted, well-built homes will adorn every farm."[2] It may have sounded visionary in 1935, but the vision is now reality. Sidney Lanier's dream for the "old hills" that "the great God turn thy fate" has finally come true. Much of the credit belongs to farmers who did adopt wise management.

Wisdom is not, however, limited to the selection of tried and proven ways; it also includes trial and use of new technology to solve old problems. Some believe technology is the cause of environmental problems. Barry Commoner (1990) said about the role of technology in improved production "These changes have turned the nation's farms, factories, vehicles, and shops into seedbeds of pollution." To some extent it is true, but those who decry our increased dependence on technology miss the point that improved technology has been the main means for recovery of the environment. One of the main ways advanced technology improved the environment was the increase in productivity of land and people. If Georgia (and the United States) had to produce food and manufactured goods for its present population with agricultural and industrial technology of the early twentieth century, our environment would be as terrible as our standard of living.

There are countless examples of technological advances, but an early one that most modern Georgians won't remember is the elimination of the screw worm fly that was so brutal to farm animals and deadly to wild ones. Rachel Carson cited the example as an alternative to insecticides in a chapter entitled "The Other Road" in *Silent Spring*.[3] The screw worm fly lays its eggs in fresh open wounds and its maggots eat the living flesh. Lack of treatment usually means death, and naturally for wild animals that was the result. The fly used to migrate from South Florida each year across Georgia infecting animals that had either deep or superficial wounds.

In the 1950s, the US Department of Agriculture hatched billions of screw worm flies, sterilized them with radiation, and released them by airplane in Florida, South Georgia, and South Alabama. The sterilized females laid eggs that would not hatch, and if the sterilized males mated with wild females, eggs of the wild females would also not hatch. By releasing these sterile flies in numbers sometimes five to ten times the wild population, screw worm flies were intercepted in their migration toward Georgia and the rest of the southeast. The last documented case of screw worm in the southeast was in 1959 and Rachel Carson heralded its extinction as "a triumphant demonstration of the worth of scientific creativity." Although the effect of this "high tech" insect control on wildlife was not documented, it came at about the time deer were beginning a comeback in Georgia and was certainly welcome if not decisive. Its effect on the health of farm animals was quick and clear to all of us who lived on farms at the time. Eradication of the screw worm in Georgia is an example of the extinction of a species that no one has complained about since.

Clearing of the air in Georgia is bound up in the technology of measuring and removing pollutant particles and gases. Modern equipment can measure concentrations of sulfur gases and ozone that were not even considered at mid-century. Technology that allows removal of pollutants from factory stacks and car exhaust makes Atlanta much more livable today than decades ago when only one-fourth or one-third as many people lived there. Particles of smoke and dirt in Atlanta's air are 75 percent less than in the 1950s, and sulfur dioxide, the main component of acid rain, is 75 percent less than in the early 1970s. Ozone has not increased since the 1970s in spite of the growth of Atlanta's highways and cars.

Cleaning of the waters of Georgia and the rest of the country is one of the great successes of environmental technology. The treatment of sewage released into streams probably has done more than any other practice to improve the health of both our streams and people. This technology did not exist at the beginning of the twentieth century and it has been continually improved since then. Industries that used to foul so much of our water have, under considerable pressure, developed technology that removes pollutants before returning the water used in manufacture. Mercury being dumped in Georgia streams has decreased by more than 90 percent, and its concentration in fish has dropped by half since the 1970s and by 60 to 80 percent in those streams which were most polluted. Improved analytical technology has allowed the detection of dangerous chemicals that fifty years ago would have gone unknown into streams.

Removal of so many of the pollutants has made water safer for humans and has allowed the return of fish to streams where they could not live in the middle of the century. As a result, fish are now more plentiful and healthier than then. Much of the improvement of Georgia's fisheries, however, lies in the tens of thousands of lakes and ponds built in the last half-century. They not only act as filters of pollutants, but have added a half million acres of habitat, not just for fish, but for many other aquatic birds and animals. Construction of these reservoirs probably means that there has been little or no net loss of wetlands in Georgia in the twentieth century.

Wild birds in twentieth-century Georgia were more plentiful at the end than any other time. Both the bald eagle and the peregrine falcon have been taken off the endangered species list, although they are still not numerous in the state. Quail, and possibly doves, have decreased, but wild turkeys, geese, ducks, and bluebirds have increased, and the number of winter bird species has tripled at Macon since 1925. Out of thirty-six birds that breed in Georgia and winter in the tropics, nine increased and only four decreased since 1966. Springtime in Georgia is not as silent now as in 1962 when Rachel Carson warned the nation about indiscriminate use of pesticides.

Other wild animals have also prospered in the last half of the twentieth century. Deer numbers have increased from about 12,000 in 1937 to more than one million in 2000. It is a rather strange coincidence that more injury and death occur in Georgia from car collisions with deer, a prominent symbol of the improved environment, than from two of the most widely studied pollutants, nitrate in drinking water and mercury in fish. (I could find no records of injury or death from either pollutant in Georgia.) Alligators have grown from endangered status to nuisance populations across South Georgia. Black bear are now numerous enough to wander into Atlanta suburbs out of the deep woods where they used to reside in low numbers. The beaver now has streams dammed all over the state where there were nearly none in the 1930s. Lenz (1994) declared that fur-bearing animal populations are "in fact healthier than they've been in a long, long time, maybe 150 years." Although nobody counts snakes, lizards, turtles, flying squirrels, or gray squirrels for that matter, their numbers have likely grown because of the increase in forests and wild areas; rural stretches abandoned during the urbanization of Georgia. Because there are so many animals uncounted, it is hard to know how the "real" balance of animals has changed during the twentieth century. However, the obvious changes in the landscape and human population indicate that forest dwellers and those that choose to be further from human habitation must have increased. The number of people living on farms in Georgia has decreased 95 percent since 1925, from 1.5 million to 80,000 in 1990. Restoration of animals so scarce in my youth, such as deer and bobcat, and those that have come to Georgia in the twentieth century, such as the coyote and armadillo, have made our animal kingdom more diverse.

Although the emphasis is most often on what is wrong with Georgia's environment, there is much more that is right with it. When the extent of progress is revealed it is amazing. It apparently is not well-known because it is seldom reported. But it is available. Get the Environmental Protection Agency's National Air Quality and Emissions Trends Reports for the 1970s to 2000 and look at the trends for air pollutant concentrations in Georgia's five or six metropolitan areas. The progress is spelled out clearly in tables of data (the older the information, the harder it is to find). Curiously, the news doesn't get further than government reports and scientific publications. The goal of this book is to make the progress obvious by bringing to light facts and trends that are buried in countless studies and documents. I hope the goal has been reached.

[1] Lanier, 1899.
[2] Sisk, 1975.
[3] Carson, 1962.

REFERENCES

Aderhold, O. C. "The Need for Conservation Education," 2:37–47. In *Proceedings of the Sixth Annual Meeting of the Georgia Chapter, Soil Conservation Society of America*, 1962.
Ager, L. M. 1991. "Changes in the sport fishery and fish population of West Point Lake, since impoundment," 154–57. In *Proceedings of the 1991 Georgia Water Resources Conference*. 19–20 March 1991. K. J. Hatcher, ed., Athens GA.
Albany Journal. 1940. "*Three-Year quail holiday.*" Conservation comments from Georgia editors. *Outdoor Georgia*, vol. 1, p. 27.
Albert, F. A. and A. H. Spector. 1955. "A New Song on the Muddy Chattahoochee," 205–10. In US Department Agriculture, Yearbook of Agriculture (*Water*).
Alexander, C., J. Ertel, R. Lee, B. Loganathan, J. Martin, R. Smith, S. Wakeham, and H. Windham. 1994. *Figures and Tables*. Volume 2 of *Pollution History of the Savannah Estuary*. Skidaway Institute of Oceanography. Savannah GA.
Allen, G. W. 1948. "The Management of Georgia Deer." *Journal of Wildlife Management*. 12:428–32.
Almand, J. D. 1965. "A Contribution to the Management Requirements of the Wood Duck (*Aix sponsa*) in the Piedmont of Georgia." MS thesis, University of Georgia, Athens GA.
Annest, J. L., K. R. Mahaffey, D. H. Cox, and J. Roberts. 1982. "Blood lead levels for persons 6 months–74 years of age: United States, 1976–80." Advance Data from Vital and Health Statistics, No. 79, US Department of Health and Human Services Publication No. 82-150.
Anonymous. 1998. "Revisions to the Georgia Regular Species list." *Oriole*, 63:102–106.
Associated Press. "Moultrie Shows How It's Done in 9-year Cleanup of the Ochlockonee River." *Atlanta Journal-Constitution*, 5 September 1989.
Associated Press. "Keep an Eye Out for Roving Deer While on the Road." *Atlanta Journal-Constitution*, 5 November 1999.
Associated Press. "Most in Poll Want Action on Harmful Conditions." *Atlanta Journal-Constitution*, 19 June 2000.
Associated Press. 2001a. "Game Officials Tranquilize Wayward Bear." *Athens Daily News/Banner-Herald*, 10 February 2001.
Associated Press. 2001b. "Shrimpers Ponder Clam Farms." *The Charleston Post and Courier*, 16 January 2001.
Atlanta Constitution. "Too Dirty for Too Long." *The Atlanta Constitution*, 17 March 1999.
Atlanta Constitution. "Clayton Wetlands Project a Clean Hit." *The Atlanta Constitution*, 18 December 2000.
Atlanta Journal. 1949. "Science Puts Finger on Bad Atlanta Air," *Atlanta Journal*, 6 May 1949.
Atlanta Journal. 1965. "Our Filthy Society." 29 March 1965.
Avery, A. A. 1999. "Infantile Methemoglobinemia: Reexamining the Role of Drinking Water Nitrates." *Environmental Health Perspectives*, 107:583–86.
Bachtel, D. C. 1990. "Georgia's Changing Agricultural Environment: An Industry in Transition." *Issues Facing Georgia*. Volume 2, No. 10. University of Georgia Cooperative Extension Service. Athens GA.

Bagby, G. T. 1969. "Our Ruined Rivers." *Georgia Game and Fish*, 7:2–16.
Bailey, M. N. 1997. "Cries in the Georgia Night." *Georgia Journal*. p. 6. September/October issue. Decatur GA.
Barnes, I. R. 1951. "The Bald Eagle: With a Price on Its Head." *Atlantic Naturalist*, 7:3–13.
Barrows, H. H., J. V. Phillips, and J. E. Brantly. 1917. "Agricultural Drainage in Georgia." Georgia Geological Survey of Bulletin No. 32.
Bartlett P.F. *American Dreams, Rural Realities. Family Farms in Crisis*. Chapel Hill NC: University of North Carolina Press, 1993.
Baskett, T. S. and M. W. Sayre. "Characteristics and importance," 1–7. In *Ecology and Management of the Mourning Dove*. T. S. Baskett, M. W. Sayre, R. E. Tomlinson, and R. E. Mirarchi, eds. Harrisburg PA: Stackpole Books, 1993.
Bates, G., G. L. Valentine, and F. H. Sprague. "Waterfowl Habitat Created by Floodwater-retarding Structures in the Southern United States," 419–25 In *Waterfowl in Winter*. M. W. Weller, ed. Minneapolis: University of Minnesota Press, 1988.
Bayne, D. R., J. M. Lawrence, and J. A. McGuire. 1983. "Primary Productivity Studies during Early Years of West Point Reservoir, Alabama-Georgia." *Freshwater Biology*, 13:477–89.
Bayne, D. R., W. C. Seesock, C. E. Webber, and J. A. McGuire. 1990. "Cultural Eutrophication of West Point Lake—A 10-year Study." *Hydrobiologia*, 199:143–56.
Bennett, H. H. *Elements of Soil Conservation*. Second edition. New York NY: McGraw-Hill Book Co., 1955.
Berger, D. D., C. R. Sindelar, Jr., and K. E. Gamble. "The Status of Breeding Peregrines in the Eastern United States", 165–73. In *Peregrine Falcon Populations, Their Biology and Decline*. J. J. Hickey, ed. Madison: University of Wisconsin Press, 1969.
Blackshaw, K. T. and J. R. Hitt. 1992. *A Birder's Guide to Georgia*. Fourth edition. Occasional Pub. No. 12, Georgia Ornithological Society, Cartersville GA.
Bledsoe, J. "Trends in Education," 171–204. In *Georgia Today: Facts and Trends*. J. C. Belcher and K. I. Dean, eds. Athens: University of Georgia Press, 1960.
Braden, B. and P. Hagan. 1989. *A Dream Takes Flight*. Atlanta: Atlanta Historical Society; Athens:University of Georgia Press, 1989.
Brender, E. V. 1952. "From Forest to Farm to Forest Again." *American Forests*, 58:24, 25, 40, 43.
Brennan, L. A., R. T. Engstrom, W. E. Palmer and S. M. Hermann G. A. Hurst, L. W. Burger, and C. L. Hardy. 1998. "Whither Wildlife Without Fire?" *Transactions of the 63rd North American Wildlife and Natural Resources Conference*. pp. 402–14.
Brimblecombe, P. *The Big Smoke: A History of Air Pollution in London Since Medieval Times*. New York NY: Methuen and Co., 1987.
Broad River Soil and Water Conservation District, Oconee River Soil and Water Conservation District, Clarke, Jackson, and Madison Counties, and the State Highway Department. 1967, 1974. "Work Plan for Little Sandy Creek and Trail Creek Watershed"; "Supplemental Watershed Work Plan #1, 1969"; "Supplemental Watershed Work Plan and Agreement # 2, 1974." Athans GA. US Department of Agriculture, Soil Conservation Service.
Brooks, R. P. 1931. *Highway Beautification Address to the Patriotic Citizens of Georgia*. Sponsored by the Atlanta Women's Club, Atlanta GA. Doris Head Brooks, 1931.
Brown, C. B. 1948. "How Effective Are Soil Conservation Measures in Sedimentation Control?," 259–65. In *Proceedings of the Federal Inter-Agency Sedimentation Conference*. Bureau of Reclamation, US Department of the Interior, Washington, DC.
Brunswig, N. L. and A. S. Johnson. 1973. "Bobwhite Quail Foods and Populations on Pine Plantations in the Georgia Piedmont during the First Seven Years Following Site Preparation." *Proceedings of the Annual Conference of the Southeastern Association of Game and Fish Commissioners*, 26:96–107.
Buell, G. R. and C. A. Couch. 1995. "National Water-quality Assessment Program: Environmental Distribution of Organochlorine Compounds in the Apalachicola—

Chattahoochee—Flint River Basin," 46–53. In *Proceedings of the 1995 Georgia Water Resources Conference*. 11–12 April 1995. K. J. Hatcher, ed. Athens GA.

Buell, G. R. and S. C. Grams. 1985. "The Hydrologic Bench-Mark Program: A Standard to Evaluate Time-Series Trends in Selected Water-Quality Constituents for Streams in Georgia." Water Resources Investigations Report 84-4318. US Geological Survey. Doraville GA.

Buie, T. S. 1937. *Proceedings of the 16th Annual Meeting of the Georgia Forestry Association*, 38–44. Athens GA.

Burke, M. A. "Historic Evolution of Channel Morphology and Riverine Wetland Hydrolic Functions in the Piedmont of Georgia." MS thesis, University of Georgia, 1996.

Burleigh, T. D. 1937. "The Birds of Athens, Clarke County, Georgia. Part I." *Oriole*, 2:11–15.

Burleigh, T. D. *Georgia Birds*. Norman OK: University of Oklahoma Press, 1958.

Burruss Institute of Public Service. 1999. "Lake Allatoona Phase I Diagnostic-Feasibility Study Report for 1992–1997." Prepared for the U.S. Environmental Protection Agency. Georgia Environmental Protection Division Contract #751-290083 and Bartow County Government, Cherokee County Water Authority, City of Cartersville, Cobb County. Government, and Cobb-Marietta Water Authority by the A. L. Burruss Institute of Public Service, Kennesaw State University, Kennesaw GA.

Bush, P. B., A. W. Tyson, R. Perkins, and W. Segars. 1997. "Results of Georgia Domestic Well Water Testing Program," 358–60. In *Proceedings of the 1997 Georgia Water Resources Conference*. 20–22 March 1997. K. J. Hatcher, ed. Athens GA.

Caldwell, I. S. *The Bunglers*. New Haven CT: The Galton Publishing Company, 1930.

Callahan, J. T., L. E. Newcomb, and J. W. Geurin. 1965. "Water in Georgia." US Geological Survey Water-Supply Paper 1762. Washington, DC.

Callender, E. and P. C. van Metre. 1997. "Reservoir Sediment Cores Show US Lead Declines." *Environmental Science and Technology*, 31:424A–28A.

Campo, C. "Erosion Woes May Doom New Projects in City." *Atlanta Journal-Constitution*, 5 June 1999.

Cappellato, R., N. E. Peters, and T. P. Meyers. 1998. "Above-ground Sulfur Cycling in Adjacent Coniferous and Deciduous Forests and Watershed Sulfur Retention in the Georgia Piedmont, U.S.A." *Water, Soil, and Air Pollution*, 103:151–71.

Carley, D. H. 1968. "Economic Analysis of the Commercial Fishery Industry of Georgia." University of Georgia, Agricultural Experiment Station Research Bulletin 37. Experiment GA.

Carson, R. *Silent Spring*. New York NY: Houghton Miflin, 1962.

Carter, J. 1973. "Sediment control in Georgia." pp. 1–4. Keynote address, Georgia Governor's Conference on Sediment Control. 22–24 July 1973. The State Soil and Water Conservation Committee. Athens GA.

———. *An Outdoor Journal*. Second edition. Fayetteville AR: University of Arkansas Press, 1994.

———. *An Hour before Daylight—Memories of a Rural Boyhood*. New York NY: Simon and Schuster, 2001.

Centers for Disease Control. 1991. "Preventing Lead Poisoning in Young Children, Statement of the National Center for Environmental Health and Injury Control." MS-F29. US Public Health Service Centers for Disease Control website www.cdc.gov/.

———. 2001. "National Report on Human Exposure to Environmental Chemicals. Lead CAS no. 7439-92-1." US Public Health Service Centers for Disease Control website www.cdc.gov/nceh/lead/lead.htm

Chalker, F. M. *Pioneer Days Along the Ocmulgee*. Carrollton GA: F. M. Chalker, 1970.

Chameides, W. L. and E. B. Cowling. 1995. *The State of the Southern Oxidants Study (SOS): Policy Relevant Findings In Ozone Pollution Research 1988–1994*. Southern Oxidants Study, College of Forest Resources, North Carolina State University, Raleigh NC.

Chameides, W. L., R. W. Lindsay, J. Richardson, and C. S. Kiang. 1988. "The Role of Biogenic Hydrocarbons in Urban Photochemical Smog: Atlanta as a Case Study." *Science*, 241:1473–75.

Chapman, D. "Area Lures 880,000 in One Decade." *Atlanta Constitution*, 30 August 2000.
Chapman, P. W., F. W. Fitch, Jr., and C. L. Veatch. *Conserving Soil Resources: A Guide to Better Living*. Atlanta GA: Turner E. Smith and Co., 1950.
Citizens Fact-Finding Movement of Georgia. *Georgia Facts in Figures. A Source Book*. Athens: University of Georgia Press, 1946.
Clark, J. D., J. H. Jenkins, P. B. Bush, and E. B. Moser. 1981. "Pollution Trends in River Otter in Georgia." *Proceedings of the Annual Conference of the Southeastern Association of Fish and Wildlife Agencies*. 35:71–79.
Clark, T. D. *The Greening of the South. The Recovery of the Land and Forest*. Lexington KY: The University of Kentucky Press, 1984.
Cofer, H. E., W. L. Tietjen, and W. J. Wysochansky. 1991. "Nutrient Contributions of a Coastal Plain Stream to Lake Blackshear," 179–82. In *Proceedings of the 1991 Georgia Water Resources Conference*. 19–20 March 1991. K. J. Hatcher, ed. Athens GA.
Cumbie, P. M. "Mercury Accumulation in Wildlife in the Southeast." Ph. D. dissertation, University of Georgia, 1975.
Collier, G. H. 1964. "The Settlement of Scull Shoals." Papers of the Athens Historical Society. 1:69–74. Athens Historical Society, Athens GA.
Commoner, B. *Making Peace with the Planet*. New York NY: Pantheon Books, 1990.
Coulter, E. M. *Old Petersburg and the Broad River Valley of Georgia: Their Rise and Decline*. Athens: University of Georgia Press, 1965.
Council on Environmental Quality. 1994. "Environmental Quality." 24th Annual Report of the Council on Environmental Quality. US Government Printing Office. Washington, DC.
———. 1997. "Environmental Quality." 25th Anniversary Report of the Council on Environmental Quality. US Government Printing Office. Washington, DC.
Cowell, C. M. 1998. "Historical Change in Vegetation and Disturbance on the Georgia Piedmont." *American Midland Naturalist*, 140:78–89.
Craft, C. B. and W. P. Casey. 1999. "Sediment and Nutrient Accumulation in Floodplain and Depressional Cypress-Gum Forest Soils of Southwestern Georgia," 443–46. In *Proceedings of the 1999 Georgia Water Resources Conference*. 30–31 March 1999. K. J. Hatcher, ed. Athens GA.
Crawford, R. L. 1998. "The Birds of Thomas County, Georgia. Revised through 1997." *Oriole*, 63:1–28.
D'Itri, P. A. and F. M. D'Itri. *Mercury Contamination: A Human Tragedy*. New York NY: John Wiley and Sons, 1977.
Dahl, T. E. and C. E. Johnson. 1991. "Status and Trends of Wetlands in the Conterminous United States, Mid-1970's to Mid-1980s." US Department of the Interior, Fish and Wildlife Service. Washington, DC.
Daniel, F. 1949. "Biologist Here to Study State's Loss of Doves." *Atlanta Journal*, 27 June 1949.
Darlington, T. L., D. F. Kahlbaum, J. M. Huess, and G. T. Wolff. 1997. "Analysis of PM-10 Trends in the United States from 1988 through 1995." *Journal of Air and Waste Management*, 47:1070–78.
Davis, S. C. 1997. "Transportation Energy Data Book." Seventeenth edition. Prepared for Office of Transportation Technologies, US Department of Energy, by the Oak Ridge National Laboratory, Oak Ridge TN.
Davis, W. H. and P. Roca. *Bluebirds and Their Survival*. Lexington: The University of Kentucky Press, 1995.
Dendy, F. E. and W. A. Champion. 1978. "Sediment Deposition in US Reservoirs. Summary of Data Reported Through 1975." US Department of Agriculture Miscellaneous Publication No. 1362. US Department of Agriculture, Washington, DC.
Denton, J. F., W. W. Baker, M. N. Hopkins Jr., L. B. Davenport, Jr., and C. S. Robbins. 1977. "Annotated Checklist of Georgia Birds." Ocassional Publication No. 6, Georgia Ornithological Society.

DeVivo, J. C., E. A. Frick, D. J. Hippe, and G. R. Buell. 1995. "National Water-quality Assessment Program: Effect of Restricted Phosphate Detergent Use and Mandated Upgrades at Two Wastewater Treatment Facilities on Water Quality, Metropolitan Atlanta, Georgia, 1988–93," 54–56. In *Proceedings of the 1995 Georgia Water Resources Conference.* April 11–12, 1995. K. J. Hatcher, ed. Athens GA.

Dolton, D. D. 1995. "Mourning Doves," 71. In *Our Living Resources; A Report to the Nation on the Distribution, Abundance, and Health of US Plants, Animals, and Ecosystems.* E. T. LaRoe, ed. US Department of the Interior, National Biological Service, Washington, DC.

Dolton, D. D. and G. W. Smith. 1998. "Mourning Dove Breeding Population Status, 1998." US Department of the Interior, US Fish and Wildlife Service. Washington, DC.

Dozier, J. (moderator). 1991. "Panel Discussion on Water Quality Regulations," 7–8. In *Proceedings of the 1991 Georgia Water Resources Conference.* 19–20 March 1991. K. J. Hatcher, ed. Athens GA.

Eakin, H. M. 1939. "Silting of Reservoirs." US Department of Agriculture Technical Bulletin No. 524.

East Georgia Planning Council. 1937. "Commercial Fisheries of Georgia." Cooperating with National Resources Committee and Works Progress Administration. Savannah GA.

Eldredge, R. L. "Peach Buzz–Buck's Head Revisited." *Atlanta Journal-Constitution,* 11 November 1999.

Elsom, D. *Atmospheric Pollution: Causes, Effects, and Control Policies.* New York NY: Basil Blackwell, Inc., 1987.

Evans, L. B., L. N. Duncan, and G. W. Duncan.. Book 5 of *Farm Life Readers.* Boston MA: Silver, Burdett and Company, 1913.

Farmers and Consumers Market Bulletin. 2001. "Provide Proper Housing for Eastern Bluebirds." Volume 84, No. 11. p. 15–16. 14 March 2001.

Faye, R. E., W. P. Carey, J. K. Stamer, and R. L. Kleckner. 1980. "Erosion, Sediment Discharge, and Channel Morphology in the Upper Chattahoochee River Basin, Georgia. With a Discussion of the Contribution of Suspended Sediment to Stream Quality." Geological Survey Professional Paper 1107. Washington, DC.

Federal Aviation Administration. 1998. "Aviation Capacity Enhancement Plan." Prepared jointly by the Federal Aviation Administration, Office of System Capacity, JIL Information Systems, and Fu Associates.

Ferguson, B. K. 1997a. "Flood and Sediment Interpretation at the Historic Scull Shoals Mill," 253–56. In *Proceedings of the 1997 Georgia Water Resources Conference.* 20–22 March 1997. K. J. Hatcher, ed. Athens GA.

———. 1997b. "The Alluvial Progress of Piedmont Streams," 132–43. In *Effects of Watershed Development and Management on Aquatic Ecosystems.* L. A. Roesner, ed. American Society of Civil Engineers, New York NY.

———. and P. W. Suckling. 1990. "Changing Rainfall-runoff Relationships in the Urbanizing Peachtree Creek Watershed, Atlanta, Georgia." *Water Resources Bulletin,* 26:313–22.

Fite, G. L. *Cotton Fields No More: Southern Agriculture 1865–1980.* Lexington: University Press of Kentucky, 1984.

Fowler, D. L. "Fish Assemblage Characteristics of Acid-Sensitive Streams in the Southern Appalachian Mountains." MS thesis, University of Georgia, 1985.

Frey, J. E. 1981. "A Fisheries Survey of the Oconee River." Georgia Department of Natural Resources, Game and Fish Division, Atlanta GA.

Frick, E. A. 1997. "Surface-water and Shallow Ground-water Quality in the Vicinity of Metropolitan Atlanta, Upper Chattahoochee River Basin, Georgia, 1992–1995," 113–17. In *Proceedings of the 1997 Georgia Water Resources Conference.* 20–22 March 1997. K. J. Hatcher, ed. Athens GA.

——— and C. A. Crandall. 1995. "National Water-quality Assessment Program: Water Quality in Surficial Aquifers in Two Agricultural Areas in Georgia, Alabama, and Florida," 42–45. In

Proceedings of the 1995 Georgia Water Resources Conference. 11–12 April 1995. K. J. Hatcher, ed. Athens GA.

———, D. J. Hippe, G. R. Buell, C. A. Couch, E. H. Hopkins, D. J. Wangsness, and J. W. Garrett. 1998. "Water Quality in the Apalachicola-Chattahoochee-Flint River Basin, Georgia, Alabama, and Florida, 1992–95." US Geological Survey Circular 1164.

———, G. R. Buell, and E. E. Hopkins. 1996. "Nutrient Sources and Analysis of Nutrient Water-Quality Data, Apalachicola-Chattahoochee-Flint River Basin, Georgia, Alabama, and Florida, 1972–90." Water Resources Investigations Report 96-4101. US Geological Survey, Atlanta GA.

Gamble, J. F. 1998. "PM2.5 and Mortality in Long-term Prospective Cohort Studies: Cause-effect or Statistical Associations?" *Environmental Health Perspectives*, 106:535–49.

Gaskin, J. W. and W. L. Nutter. 1989. "The Effect of Irrigation with Pretreated Wastewater on Groundwater Quality and Elevation at Clayton County, Georgia," 207–209. In *Proceedings of the 1989 Georgia Water Resources Conference.* 16–17 May 1989. K. J. Hatcher, ed. Athens GA.

Geihsler, M. R. and D. R. Holder. 1980. "Balanced Success and an Evaluation of Factors Associated with Balance Success in Newly Stocked Georgia Ponds." Final Report Federal Aid Project G-1. Georgia Department of Natural Resources, Game and Fish Division, Atlanta GA.

Geldert, L. N., ed. *Facts about Georgia.* Atlanta GA: Foote and Davies Company, 1916.

Georgia Agricultural Statistics Service. 1950–2000. "Georgia Agricultural facts." Georgia Department of Agriculture in Cooperation with the US Department of Agriculture, National Agricultural Statistics Service. Athens GA.

The Georgia Conservancy. *Air Pollution in Savannah. A Report on the State of the Air in Chatham County, Georgia and What Can Be Done About It.* Savannah GA: The Georgia Conservancy, 1979.

Georgia Department of Agriculture. *Georgia: Historical and Industrial.* Atlanta GA: Franklin Printing and Publishing Company, 1901.

Georgia Department of Community Affairs. 1997. "Georgia Solid Waste Management Plan." Georgia Department of Community Affairs, Georgia Department of Natural Resources, and Georgia Environmental Facilities Authority, Atlanta GA.

Georgia Department of Game and Fish. -1926. *Birds on Georgia Farms.* Southeastern Printing Company, Atlanta GA.

Georgia Department of Natural Resources. 1938. 1937–1938 Biennial Report. By R. F. Burch.

———. 1976. "Water Quality Control in Georgia, 1975." Environmental Protection Division, Georgia Department of Natural Resources. Atlanta GA.

———. 1985–2000. "Annual Report of Air Pollution Measurements of the Georgia Air Quality Surveillance Network." Environmental Protection Division, Air Protection Branch, Atlanta GA.

———. 1988. "Annual Report." Atlanta GA.

———. 1990a. "Decades of Challenge: Georgia's Environment 1964 to 1990." Georgia Department of Natural Resources, Atlanta GA.

———. 1990b. "Jackson Lake Phase I Diagnostic/Feasibility Study Final Report." Georgia Department of Natural Resources, Environmental Protection Division. Atlanta GA.

———. 1993. "Trends in Georgia Wildlife, 1974–1993." Wildlife Resources Division, Social Circle GA.

———. 1993–1999. "Georgia Wildlife Surveys." Wildlife Resources Division, Social Circle GA.

———. 1996a. "State of Georgia Landcover Statistics by County." Project Report 26. Georgia Natural Heritage Program, Social Circle GA.

———. 1996b. "Ambient Air Surveillance Report." Environmental Protection Division, Air Protection Branch, Atlanta GA.

———. 1997. "Water Quality in Georgia–1994–1995." Georgia Department of Natural Resources, Environmental Protection Division. Atlanta GA.

———. 1998a. "Fish Tissue Assessment Project. Data Summary." Wildlife Resources Division, Environmental Protection Division. 1991–1997.

———. 1998b. "Record Number of Deer Donated to Hunters for the Hungry During the First Collection Weekend." Press Release of the Georgia Wildlife Resources Division, 13 November 1998. From the Georgia Wildlife Resources Division website www.ganet.org/dnr/wild/.

———. 1998c. "Georgia Offers Many Opportunities for Hunting Dove on Wildlife Management Areas." Press Release of the Georgia Wildlife Resources Division, 18 August 1998. From the Georgia Wildlife Resources Division website www.ganet.org/dnr/wild/.

———. 1998d. "Wildlife License Plate Sales Still Soaring in 1998." Press Release of the Georgia Wildlife Resources Division, 16 October 1998. From the Georgia Wildlife Resources Division website www.ganet.org/dnr/wild/.

———. 1999a. "Water Quality in Georgia—1996–1997." Georgia Department of Natural Resources, Environmental Protection Division. Atlanta GA.

———. 1999b. "Georgia's Environment 98." Georgia Department of Natural Resources, Environmental Protection Division, Atlanta GA.

———. 1999c. "State of Georgia Acquires Wood Stork Colony." Press Release of the Georgia Wildlife Resources Division, 9 July 1999. From the Georgia Wildlife Resources Division website www.ganet.org/dnr/wild/.

———. 2000a. "2000 Bald Eagle Population Survey Results: More Nests Mean Good News for Georgia's Eagles." Press Release of the Georgia Wildlife Resources Division, 16 June 2000. From the Georgia Wildlife Resources Division website www.ganet.org/dnr/wild/.

———. 2000b. "H.B. 1465 May Increase Bag Limit for Deer in Georgia." Press Release of the Georgia Wildlife Resources Division, March 6, 2000. From the Georgia Wildlife Resources Division website www.ganet.org/dnr/wild/.

———. 2000c. "WRD Predicts Good Turkey Hunting for the 2000 Spring Season." Press Release of the Georgia Wildlife Resources Division, 10 March 2000. From the Georgia Wildlife Resources Division website www.ganet.org/dnr/wild/.

———. 2000d. "DNR Biologists Working with Private Landowners to Create Important Habitat for Bobwhite Quail." Press Release of the Georgia Wildlife Resources Division, 21 May 2000. From the Georgia Wildlife Resources Division website www.ganet.org/dnr/wild/.

———. 2000e. "Purple Songsters Flying South of the Border." Press Release of the Georgia Wildlife Resources Division, 10 March 2000. From the Georgia Wildlife Resources Division website www.ganet.org/dnr/wild/.

———. 2001a. "Water Quality in Georgia—1998–1999." Georgia Department of Natural Resources website www.Georgisnet.org/dur/environ/.

———. 2001b. "Warm Weather Brings Frequent Bear Sightings to the State." Press Release of the Georgia Wildlife Resources Division, 20 April 2001. From the Georgia Wildlife Resources Division website www.ganet.org/dnr/wild/.

———. 2001c. "Georgia's Environment." Georgia Department of Natural Resources, Environmental Protection Division, Atlanta GA. From Georgia Department of Natural Resources' website www.Georgianet.org/dur/environ/.

Georgia Forestry Commission. 1946–1999. "Annual Reports and Monthly Fire Reports." Georgia Forestry Commission, Macon GA.

Georgia Game and Fish. 1963. "Oconee Population Study Shows Game Fish Collapse." *Georgia Game and Fish*, 1:11, 24.

———. 1969. "Save the Alcovy and Lake Jackson." Editorial, *Georgia Game and Fish*, 4(6) Back of cover and p. 15. June 1969.

Georgia Geological Survey. 1985–2001. "Ground-water Quality [and Availability] in Georgia for [1984 to 1999]." Georgia Geological Survey Circular 12A through 120. Georgia Department of Natural Resources, Atlanta GA.

"Georgia Nonpoint Source Management Plan." 1989. Georgia Department of Natural Resources, Environmental Protection Division. Atlanta GA.

Georgia Ornithological Society. 2000. "91st–99th National Audubon Society Christmas Bird Count Results." From the Georgia Ornithological Society website www.gos.org.

Georgia State Soil and Water Conservation Needs Inventory Committee. 1962. "Georgia Soil and Water Conservation Needs Inventory." US Department of Agriculture, University of Georgia, College of Agriculture, and The State of Georgia.

Georgia Statistical Abstract. Various years. L. M. Akioka, ed. Terry College of Business, University of Georgia, Athens GA.

Georgia Water Quality Control Board. 1971. "Mercury Pollution Investigation in Georgia, 1970–1971." Atlanta GA.

Georgia Water Use and Conservation Committee. 1955. *Water in Georgia.* Printed by the Georgia Water Law Revision Commission.

Geron, C. D., T. E. Pierce, and A. B. Guenther. 1995. "Reassessment of Biogenic Volatile Organic Compound Emissions in the Atlanta Area." *Atmospheric Environment,* 29:1569–78.

Giddens, J. 1975. "Contamination of Water by Air Pollutants, Especially Ammonia from Animal Manures." Completion Report, USDI/OWRT project No. A-050-GA. Department of Agronomy, University of Georgia, Athens GA in cooperation with the Environmental Resources Center, Georgia Institute of Technology, Atlanta GA.

Gill, F. 1993. "The Bald Eagle." In *The Birds of North America,* A. Poole and F. Gill, eds. The Academy of Natural Sciences; Washington DC: The American Ornithologists' Union PA.

Glenn, L. C. 1911. "Denudation and Erosion in the Southern Appalachian Region and the Monongahela Basin." US Geological Survey Professional Paper 72. Washington, DC.

Gobris, N. M. "Habitant Occupancy during the Breeding Season by Bachman's Sparrow at Piedmont National Wildlife Refuge in Central Georgia." MS thesis, University of Georgia, 1992.

Godbee, J. F., Jr. "The ESA and the Wood Stork: Private Landowner Implications," 64–68. In *Peregrine Falcon Populations, Their Biology and Decline.* J. J. Hickey, ed. Madison WI: The University of Wisconsin Press, 1993.

Gordon, S., L. R. John, R. F. Smith, A. S. Hazzard, H. T. Lee, R. F. Poston, and I. T. Quinby. 1964. "Management and Operations of the Georgia Game and Fish Department." A Report to the Georgia Game and Fish Commission in cooperation with the Governor's Commission for Efficiency and Improvement in Government. Atlanta GA.

Gould, M. C. 1995. "Nitrate-nitrogen Levels in Well Water from the Little River/Rooty Creek Watershed in East Central Georgia," 148–51. In *Proceedings of the 1995 Georgia Water Resources Conference.* 11–12 April 1995. K. J. Hatcher, ed. Athens GA.

Graham, F. Jr. 1990. "2001:_Birds That Won't Be with Us." *American Birds,* 44:1074–81, 1194–99.

Green, C. H. *Birds of the South.* Chapel Hill NC: University of North Carolina Press, 1933.

Greenberg, M. R. "Health and Risk in Urban-industrial Society: An Introduction," 3–24. In *Public Health and the Environment.* M. R. Greenberg, ed. New York NY: The Guilford Press, 1987.

Greene, E. R., W. W. Griffin, E. P. Odum, H. L. Stoddard, and I. R. Tompkins. 1945. "Birds of Georgia: A Preliminary Check-List and Bibliography of Georgia Ornithology." Occasional Paper No. 2, Georgia Ornithological Society. University of Georgia Press. Athens GA.

Gschwandtner, G., K. C. Gschwandtner, and K. Eldridge. 1985. "Historic Emissions of Sulfur and Nitrogen Oxides in the United States from 1900 to 1980." Project Summary. US Environmental Protection Agency, Air and Engineering Research Laboratory. Research Triangle Park NC.

Haines, B. 1979. "Acid Precipitation in Southeastern United States: A Brief Review." *Georgia Journal of Science,* 37:185–91.

Halbrook, R. S., J. H. Jenkins, P. B. Bush, and N. D. Seabolt. 1994. "Sublethal Concentrations of Mercury in River Otters: Monitoring Environmental Contamination." *Archives of Environmental Contamination Toxicology,* 27:306–10.

Hall, J. A. and W. Rogers. 1928. *Some Helpful Georgia Birds*. State Board of Game and Fish. Atlanta GA.

Ham, L. K. and H. H. Hatzell. 1996. "Analysis of Nutrients in the Surface Waters of the Georgia-Florida Coastal Plain Study Unit, 1970–91." US Geological Survey: Water-Resources Investigations Report 96-4037. National Water Quality Assessment Program. Tallahassee FL.

Hardie, A. "State Tries to Pump Up Quail Population for Hunting." *Atlanta Journal-Constitution*, 26 April 1999.

Harmon, J. "Chemical Release Scorches Area around McCaysville." *Atlanta Constitution*, 22 August 1990.

Harper, R. M. 1923. "Development of Agriculture in Georgia from 1850 to 1920." A series of four articles reprinted by the *Georgia Historical Quarterly*.

Harris, M. J. "Status of the Wood Stork in Georgia, 1965–1993," 34–46. In *Proceedings of the Wood Stork Symposium*. Savannah GA: The Georgia Conservancy, 1993.

Harris, L. D. and P. C. Frederick. "The Role of the Endangered Species Act in the Conservation of Biological Diversity: An Assessment," 99–115. In *Integrated Environmental Management*. J. Cairns Jr. and T. V. Crawford, eds. Chelsea MI: Lewis Pulishers, Inc., 1991.

Hartman, W. A. and H. H. Wooten. 1935. "Georgia Land Use Problems." Georgia Experiment Bulletin No. 191. University of Georgia, Athens GA.

Hastings, C. E. and J. E. Frey 1962. "Fisheries Survey of Georgia." Georgia Game and Fish Commission, Final Report, Fed. Aid Proj. F-7-R.

Hefner, J. M. and J. D. Brown. 1985. "Wetland Trends in the Southeastern United States." *Wetlands*, 4:1–11.

Hefner, J. M., B. O. Wilen, T. E. Dahl, and W. E. Frayer. 1994. "Southeast Wetlands–Status and Trends, Mid-1970s to Mid-1980s." US Fish and Wildlife Service. Atlanta GA.

Heinsohn, R. J. and R. L. Kabel. *Sources and Control of Air Pollution*. Upper Saddle River NJ: Prentice-Hall Inc., 1999.

Held, R. B. and M. Clawson. *Soil Conservation in Perspective*. Baltimore MD: The Johns Hopkins Press, 1965.

Hicks, T. 1977. "Beaver and Their Control in Georgia." Technical Bulletin WL 2. Georgia Department of Natural Resources. Atlanta GA.

Hilgard, E. W. 1884. "Report on Cotton Production in the United States." Department of the Interior, Census Office. Part II, pp. 502–503. US Government Printing Office, Washington, DC.

Hill, E. P., P. W. Sumner, and J. B. Wooding. 1987. "Human Influences on Range Expansion of Coyotes in the Southeast." *Wildlife Society Bulletin*, 15:521–24.

Hippe, D. J., H. H. Hatzell, L. K. Ham, and P. S. Hardy. 1995. "National Water-quality Assessment Program: Pesticide Occurrence and Temporal Distribution in Streams Draining Urban and Agricultural Basins in Georgia and Florida, 1993–94," 34–41. In *Proceedings of the 1995 Georgia Water Resources Conference*. 11–12 April 1995. K. J. Hatcher, ed. Athens GA.

Hoke, J. T. Jr. 2000. J. "Strom Thurmond Lake Analysis and Summary of Sediment Range Survey." US Army Corps of Engineers, Hydrology and Hydraulics Branch. From the Army Corps of Engineers webpage http://water.sas.usace.army.mil/thurmondsediment.

Holbrook, S. H. *Burning an Empire: The Story of American Forest Fires*. New York NY: The MacMillan Company, 1943.

Hon, T. 1979. "Relative Abundance of Bobcats in Georgia: Survey Techniques and Preliminary Results," 104–106. In *Bobcat Research Conference Proceedings. Current Research on Biology and Management of Lynx rufus*. National Wildlife Federation. Science and Technology Series 6.

Hopkins, M. N. and E. P. Odum. 1953. "Some Aspects of the Population Ecology of Breeding Mourning Doves in Georgia." *Journal of Wildlife Management*, 17:132–43.

Hopkins, M. N., Jr. 1958. "Land Usage and Birds." *Oriole*, 23:1–2.

———. 1975. "The Birdlife of Ben Hill County, Georgia and Adjacent Areas." Georgia Ornithological Society, Occasional Publication No. 5.

Hubbard, R. K., J. M. Sheridan, and L. R. Marti. 1990. "Dissolved and Suspended Solids Transport from Coastal Plain Watersheds." *Journal of Environmental Quality*, 19:413–20.

Huntington, T. G. 1996. "Assessment of the Potential Role of Atmospheric Acidic Deposition in the Pattern of Southern Pine Beetle in the Northwestern Coastal Plain of Georgia, 1992–1995." US Geological Survey Water Resources Investigations Report 96-4131. US Geological Survey and Georgia Forestry Commission, Atlanta GA.

Husar, R. B. 1986. "Emissions of Sulfur Dioxide and Nitrous Oxides and Trends for Eastern North America," 48–92 In *Acid Deposition-Long-Term Trends*. National Academy Press. Washington, DC.

——— and W. E. Wilson. 1993. "Haze and Sulfur Emission Trends in the Eastern United States." *Environmental Science and Technology*, 27:12–16.

———, J. M. Holloway, and D. E. Patterson. 1981. "Spatial and Temporal Pattern of Eastern US Haziness: A Summary." *Atmospheric Environment*, 15:1919–28.

Jackson, H. R. 1850. *Tallulah, and Other Poems*. John M. Cooper. Savannah GA.

Jackson, J. A. 1995. "The Red-cockaded Woodpecker: Two Hundred Years of Knowledge, Twenty Years under the Endangered Species Act." pp. 42–48 In *Red-cockaded Woodpecker: Recovery, Ecology and Management*. D. L. Kulhavy, R. G. Hooper, and R. Costa, eds. Center for Applied Studies in Forestry, College of Forestry, Stephen F. Austin State University, Nacogdoches TX.

Jacobs, J. *Bass Fishing in Georgia*. Atlanta GA: Peachtree Publishers, Ltd., 1998.

Jenkins, J. H. 1952. "The Extirpation and Restoration of North Georgia deer—A Sixty-year History," 472–46. In *Transactions of the Seventeenth North American Wildlife Conference*. 17–19 March 1952. Wildlife Management Institute, Washington, DC.

Jenkins, J. H. *The Game Resources of Georgia*. Atlanta GA: Georgia Game and Fish Commission, 1953.

Jenkins, J. H. 1961. "The Role of Wildlife in Soil and Water Conservation." In *Proceedings of the Fifth Annual Meeting of the Georgia Chapter, Soil Conservation Society of America*. 26–27 May 1961.

Johnson, F. A. 1948. Upland Game Management. Wildlife Management Institute. Washington, DC.

Johnson, J. 2000. Personal Communication. US Department of Agriculture, National Agricultural Statistics Service, Washington DC.

Johnson, S. C. and J. E. Kennamer. 1975. "Reproductive Success of the Resident Canada Goose Flock at the Eufaula National Wildlife Refuge." In *Proceedings of the Annual Conference of the Southeastern Association of Game Fish Commissioners* 28:617–26.

Johnson, T. G, A. Jenkins, and J. L. Wells. 1997. "Georgia's Timber Industry: An Assesment of Timber Product Output and Use, 1995." Resource Bulletin SRS-14. US Department of Agriculture Forest service, Southern ResearchStation, Asheville NC.

Johnston, D. W. and E. P. Odum. 1956. "Breeding Bird Populations in Relation to Plant Succession on the Georgia Piedmont." *Ecology*, 37:50–62.

Jordan, J. L. and A. H. Elnagheeb. 1993. "Willingness to Pay for Improvements in Drinking Water Quality." *Water Resources Research*, 29:237–45.

Kamps, D. M. 1989. "Jackson Lake: Response to Nutrient Reduction," 225–26 In *Proceedings of the 1989 Georgia Water Resources Conference*. 16–17 May 1989. K. J. Hatcher, ed. Athens GA.

King, T. G. "Response of Bird Communities to Dormant-Season Versus Growing-Season Prescribed Fire in Mature Pine Stands." M. S. thesis, University of Georgia, 1997.

Knight Ridder Newspapers. "Bald Eagles to Be Removed from Endangered List." *Milledgeville Union-Recorder*, 17 June 1999.

Koutrakis, P. and C. Sioutas. 1996. "Concentrations and Health Effects." pp. 15–39. In *Particles in Our Air*. R. Wilson and J. Spengler, eds. Cambridge MA: Harvard School of Public Health, 1996.

Kuhn, C. M., H. E. Joye, and E. B. West. *Living Atlanta. An Oral History of the City, 1914–1948.* Atlanta: The Atlanta Historical Society Atlanta; Athens: University of Georgia Press, 1990.

Kundell, J. E. 1984. "Acid Rain: The Georgia Situation." Carl Vinson Institute of Government, University of Georgia, Athens GA.

———. 1989. "Municipal Solid Waste Management in Georgia: Policy Alternatives." Carl Vinson Institute of Government, University of Georgia, Athens GA.

——— 1996. "Environmental Trends: Implications for Small Communities in the South." Carl Vinson Institute of Government, University of Georgia, Athens GA.

——— and G. Swanson. 1991. "Georgia's Air Quality: Status and Implications of the Federal Clean Air Act of 1990." Carl Vinson Institute of Government, University of Georgia, Athens GA.

——— and M. F. Dorfman. 1994. "Georgia's Environmental Policy: New Directions." Carl Vinson Institute of Government, University of Georgia, Athens GA.

——— and S. W. Woolf. 1986. "Georgia Wetlands: Trends and Policy Options." Carl Vinson Institute of Government. University of Georgia, Athens GA.

Laerm, J. 1994. "Northern Exposure. The Georgia Otter Extends Its Territory." *Georgia Wildlife,* 3(4):38–43.

Landau, E. 1967. "Economic Aspects of Air Pollution as It Relates to Agriculture," 113–26. In *Agriculture and the Quality of Our Environment.* N. C. Brady, ed. American Association for the Advancement of Science. Publication No. 85. Washington, DC.

Landers, J. L. and B. S. Mueller. 1986. "Bobwhite Quail Management: A Habitat Approach." Miscellaneous Publication No. 6 of the Tall Timbers Research Station in cooperation with Quail Unlimited. Tallahassee FL.

Langdale, G. W., J. E. Box, Jr., R. A. Leonard, and W. G. Fleming. 1979. "Corn Yield Reduction on Eroded Southern Piedmont soils." *Journal of Soil and Water Conservation,* 34:226–28.

Lanier, M. D., ed. *Poems of Sidney Lanier.* Athens GA: University of Georgia Press, 1981.

Lanier, S. *Retrospects and Prospects. Descriptive and Historical Essays.* Second edition. New York NY: Charles Scribner's Sons, 1899.

Leigh, D. S. 1997. "Mercury-tainted Overbank Sediment from Past Gold Mining in North Georgia, USA." *Environmental Geology,* 30:244–51.

Lenz, R. J. 1994. "Trappers and Furbearers in Georgia." *Georgia Wildlife,* 4(1):60–64.

Lewis, S. 2001. "Trappers Zone Out Beaver Habitats in Peachtree City." *Atlanta Constitution,* 28 May 2001.

Lindberg, S. E. and G. M Lovett. 1992. "Deposition and Forest Canopy Interactions of Airbourne Sulfur: Results from the Integrated Forest Study." *Atmospheric Environment,* 26A:1477–92.

Linton, T. L. 196?. "Feasibility Study of Methods for Improving Oyster Production in Georgia; Completion Report 2-10-R." University of Georgia, Marine Institute, Sapelo Island. Georgia Marine Fisheries Division, Atlanta GA.

Lynch, J. A., J. W. Grimm, and V. C. Bowersox. 1995. "Trends in Precipitation Chemistry in the United States: A National Perspective, 1980–1992." *Atmospheric Environment,* 29:1231–46.

Lynn, D. A., G. L. Deane, R. C. Galkiewicz, and R. M. Bradway. 1976. "National Assessment of the Urban Particulate Problem." Volume I. US Environmental Protection Agency. Office of Air and Waste Management, and Office of Air Quality Planning and Standards. Research Triangle Park NC.

Madson, J. *The Mourning Dove.* East Alton IL: Winchester Press, 1978.

Magilligan, F. J. and M. L. Stamp. 1997. "Historical Land-cover Changes and Hydromorphic Adjustment in a Small Georgia Watershed." *Annals of the Association of American Geographers,* 87:614–35.

Martin, E. M., P. H. Geissler, and A. N. Novara. 1991, 1992, 1993, 1994. "Preliminary Estimates of Waterfowl Harvest and Hunter Activity in the United States During the 1990,

1991, 1992, 1993. Hunting Season." Administrative Report of the US Department of the Interior, Fish and Wildlife Service, Washington, DC.

Martin, F. W. and J. R. Sauer. "Population Characteristics and Trends in the Eastern Management Unit," 281–304. In *Ecology and Management of the Mourning Dove.* T. S. Baskett, M. W. Sayre, R. E. Tomlinson, and R. E. Mirarchi, eds. Harrisburg PA: Stackpole Books, 1993.

Mauldin, A. C. II, and J. C. McCollum. 1992. "Status of the Chattahoochee River Fish Population Downstream of Atlanta, Georgia." Georgia Department of Natural Resources, Game and Fish Division, Final Report, Federal Aid Project F-26.

Mayhew, E. A. and M. C. Mayhew. 1993. "Adsorption and Sedimentation of Phosphorus by Clay Soils in Lake Lanier," 104–106. In *Proceedings of the 1993 Georgia Water Resources Conference.* 20–21 April 1993. K. J. Hatcher, ed. Athens GA.

McCallie, S. W. 1911. "The Drainage Situation in Georgia. In A Preliminary Report on Drainage Reclamation in Georgia." Georgia Geological Survey Bulletin No. 25. Georgia Geological Survey and US Department of Agriculture.

McCort, W. D. and M. C. Coulter. "Endangered Species Protection—The Wood Stork Example," 119–36. In *Integrated Environmental Management.* J. Cairns Jr. and T. V. Crawford, eds. Chelsea MI: Lewis Pulishers, Inc., 1991.

McCrary, J. L. and J. E. Kundell. 1997. "Georgia's Threatened Lands: The Impacts of Sprawl." A report prepared for the Turner Foundation. Carl Vinson Institute of Government, University of Georgia, Athens GA.

Meade, R. H. 1976. "Sediment problems in the Savannah River basin." pp. 105–29 In *The Future of the Savannah River. Proceedings of a Symposium.* B. L. Dillman and J. M. Stepp, eds. Water Resources Research Institute, Clemson University, Clemson SC.

———. 1982. "Sources, Sinks, and Storage of River Sediment in the Atlantic Drainage of the United States." *Journal of Geology,* 90:235–52.

Meine, C. *Aldo Leopold, His Life and Work.* Madison WI: University of Wisconsin Press, 1988.

Mellinger-Birdsong, A. K., K. E. Powell, and T. Iatridis. 2000. Georgia Asthma Report. Publication No. DPH00.65H. Georgia Department of Human Resources, Division of Public Health. Atlanta, Ga.

Merrill, M. D., M. C. Freeman, B. J. Freeman, E. A. Kramer, and L. M. Hartle. 2001. "Stream Loss and Fragmentation Due to Impoundments in the Upper Oconee Watershed." In *Proceedings of the 2001 Georgia Water Resources Conference.* 26–27 March 2001, K. J. Hatcher, ed. Athens GA.

Minor, E. "South Georgia Trees Boost Top Industry." *Athens Daily News/ Banner-Herald,* 28 February 1999.

Moffat, A. S. 1998. "Temperate Forest Gain Ground." *Science,* 282:1253. 13 November, 1998.

Montgomery, B. "The Buck Stops Here: Deer Subdued in Store." *Atlanta Journal-Constitution,* 22 February 2001.

Moran, M. *Tincture of Time. The Story of 150 years of Medicine in Atlanta. 1845 to 1995.* Atlanta GA: Williams Printing, 1995.

Mulholland, J. A., A. J. Butler, J. G. Wilkinson, A. G. Russell, and P. E. Tolbert. 1998. "Temporal and Spatial Distributions of Ozone in Atlanta: Regulatory and Epidemiologic Implications." *Journal of Air and Waste Management Association,* 48:418–26.

National Atmospheric Deposition Program/National Trends Network. (2001). NADP Program Office, Illinois State Water Survey, 2204 Griffith Drive, Champaign IL. Website http://nadp.sws.uiuc.edu/madpdate/.

National Clean Air Coalition and Friends of the Earth Foundation. 1984. "Acid Rain in the South: Its Impact and Its Threat." National Clean Air Coalition, Washington, DC; Friends of the Earth Foundation, San Francisco CA.

National Marine Fisheries Service. 2001. Personal Communications from National Marine Fisheries Service, Fisheries Statistics and Economics Division. Silver Spring MD.

National Research Council. 1986. "Summary and Synthesis," 1–47. In Acid Deposition-Long-Term Trends. National Academy Press. Washington, DC.

Oberle, M. W. and J. C. Haney. 1997. Possible Breeding Range Extensions of Northern Forest Birds in Northeast Georgia. *Oriole* 62:35–44.

Odum, E. P. 1963. "Concluding Remarks of the Co-chairman." Tall Timbers Fire Ecology Conference 2:177–80. Tall Timbers Research Station, Tallahassee.

———. and M. G. Turner. 1987. "The Georgia Landscape: A Changing Resource." Final Report of the Kellogg Physical Resources Task Force. The Institute of Ecology, University of Georgia, Athens GA.

———, O. S. Allen III, and H. R. Pulliam. 1993. "Southward Extension of Breeding Ranges of Passerine Birds in the Georgia Piedmont in Relation to the Reversed Latitudinal Gradient." *Georgia Journal of Science*, 51:131–40.

Olmsted, R. O., T. Levin, and R. Stapp. 1970. "Georgia Shellfish Sanitation Program Review, 1969–1970." Public Health Service, Food and Drug Administration, Atlanta GA.

Oppenheim, J. A. "Sedimentation Rates and Fluvial Response to Land-Use Changes in a Small Georgia Piedmont Watershed," M. A. thesis, University of Georgia, 1996.

Osowski, S. L., L. W. Brewer, O. E. Baker, and G. P. Cobb. 1995. "The Decline of Mink in Georgia, North Carolina, and South Carolina: The Role of Contaminants." *Archives of Environmental Contamination Toxicology*, 29:418–23.

Patrick, R., F. Douglas, D. M. Palavage, and P. M. Stewart. *Surface Water Quality: Have the Laws Been Successful?* Princeton: Princeton University Press, 1992.

Peterjohn, B. J., J. R. Sauer, and S. Orsillo. 1995. "Breeding Bird Survey: Population Trends 1966–92," 17–21. In *Our Living Resources; A Report to the Nation on the Distribution, Abundance, and Health of US Plants, Animals, and Ecosystems*. E. T. LaRoe, ed. US Department of the Interior, National Biological Service, Washington, DC.

Peters, N. E. and S. J. Kandell. 1997. "Evaluation of Streamwater Quality in the Atlanta Region," 94–98. In *Proceedings of the 1997 Georgia Water Resources Conference*. 20–22 March 1997. K. J. Hatcher, ed. Athens GA.

———, G. R. Buell, and E. A. Frick. 1997. "Spatial and Temporal Variability in Nutrient Concentrations in Surface Waters of the Chattahoochee River Basin near Atlanta, Georgia," 103–12. In *Proceedings of the 1997 Georgia Water Resources Conference*. 20–22 March 1997. K. J. Hatcher, ed. Athens GA.

Peterson, R. T. *Audubon Birds of America*. New York NY: Crown Publishers, Inc., 1979.

Pirkle, J. L., D. J. Brody, E. W. Gunter, R. A. Kramer, D. C. Paschal, K. M. Flegal, and T. D. Matte. 1994. "The Decline in Blood Levels in the United States. The National Health and Nutrition Examination Surveys (NHANES)." *Journal of the American Medical Association*, 272:284–91.

———, R. B. Kaufmann, D. J. Brody, T. Hickman, E. W. Gunter, and D. C. Paschal. 1998. "Exposure of the US Population to Lead, 1991–1994." *Environmental Health Perspectives*, 106:745–50.

Plummer, D. "Something to Quack About." *Atlanta Journal-Constitution*, 23 July 1997.

President's Commission on the Health Needs of the Nation. 1951. "America's Health Status, Needs, and Resources." Building America's Health. A report of the President. US Government Printing Office, Washington, DC.

Prunty, M. C. 1965. "Some Geographic Views on the Role of Fire in Settlement Processes in the South." Tall Timbers Fire Ecology Conference. 4:161–68. Tall Timbers Research Station, Tallahassee.

——— and C. S. Aiken. 1972. "The Demise of the Piedmont Cotton Region." *Annals of the Association of American Geographers*. 62:283–306.

Pullen, T. M. "Some Effects of Beaver(*Castor canadensis*) and Beaver Pond Management on the Ecology and Utilization of Fish Populations Along Warm-Water Streams in Georgia and South Carolina." Ph.D. Dissertation, University of Georgia, 1971.

Pullis, G. and A. Laughland. 1999. "Black Bass Fishing in the U.S." Addendum to the 1996 National Survey of Fishing, Hunting and Wildlife-Associated Recreation. Report 96-3. US Fish and Wildlife Service, Washington, DC.

Quertermus, C. 1996-1999. "Tournament Creel Reports," Georgia BASS Chapter Federation. State University of West Georgia, Carrollton GA.

———. 1998. "Evaluation of Fishing Quality and Relative Abundance of Black Bass (*Micropterus*) Populations at Twelve Georgia Mainstream Reservoirs, 1978-1995, Using Catch Data from Bass Club Tournaments." Georgia Department of Natural Resources, Wildlife Resources Division, Social Circle GA.

Range, W. *A Century of Georgia Agriculture, 1850-1950.* Athens GA: University of Georgia Press, 1954.

Rasmussen, T., S. Holmbeck-Pelhama, K. Baerb, and T. Laidlaw. 1998. Chapter 6: "Lake Lanier Water Quality Targets." In *Diagnostic/Feasibility Study of Lake Sidney Lanier, Georgia. Project Completion Report.* Prepared for the Georgia Environmental Protection Division under the US Environmental Protection Agency's Clean Lakes Program, June 1998.

Reeves, H. M. and R. E. McCabe. "Historical Perspective," 7-46. In *Ecology and Management of the Mourning Dove.* T. S. Baskett, M. W. Sayre, R. E. Tomlinson, and R. E. Mirarchi, eds. Harrisburg PA: Stackpole Books, 1993.

Reinolds, C. "High Pollution Levels Threaten Lake Allatoona." *Atlanta Journal-Constitution,* 5 February 2000.

Reynolds, R. V. and A. H. Pierson. 1942. "Fuel Wood Used in the United States 1630-1930." Circular 641. US Department of Agriculture, Washington, DC.

Richardson, E. C. and E. G. Diseker. 1961. "Control of Roadbank Erosion in the Southern Piedmont." *Agronomy Journal.* 53:292-94.

Richmond County Department of Health. 1964. "An Environmental Health Report of Augusta, Georgia." Sponsored by Richmond County Department of Health, The City of Augusta, and Richmond County. 3-8 Augusta 1964.

Riddleberger, K. A. and R. R. Odom. 1987. "The Status of the Osprey in Georgia." In *Proceedings of the Southeastern Nongame Endangered Wildlife Symposium.* 3:40-47.

Rivers, W. and T. Loggins. 1989. "Georgia's Fourth Forest." Georgia Forestry Commission. Macon GA.

Robbins, C. S. "Non-native Birds," 437-40. In *Our Living Resources; A Report to the Nation on the Distribution, Abundance, and Health of US Plants, Animals, and Ecosystems.* E. T. LaRoe, ed. US Department of the Interior, National Biological Service, Washington, DC, 1995.

Rostlund, E. 1957. "The Myth of a Natural Prairie Belt in Alabama: An Interpretation of Historical Records." *Annals of the Association of American Geographers* 47:392-411.

———. 1960. "The Geographic Range of the Historic Bison in the Southeast." *Annals of the Association of American Geographers.* 50:395-407.

Rowalt, E. M. 1937. "Soil Defense in the Piedmont." US Department of Agriculture Farmer's Bulletin 1767.

Ruckel, S. W. 1987. "The Status of the American Alligator in Georgia." *Proceedings of the Southeastern Nongame Endangered Wildlife Symposium.* 3:91-97.

Rykiel, E. J. Jr. "The Okefenokee Swamp Watershed: Water Balance and Nutrient Budgets." Ph. D. Dissertation, University Georgia, 1977.

Sauer, J. R., K. Pardieck, J. E. Hines, and J. Fallon. 2001. "The North American Breeding Bird Survey, Results and Analysis 1966-2000." Version 2001.1, USGS Patuxent Wildlife Research Center, Laurel MD.

———, S. Schwartz, and B. Hoover. 1996. The Christmas Bird Count Home Page. Version 95.1. Patuxent Wildlife Research Center, Laurel MD.

Saunders, F. B. "The Revolution in Agriculture," 267-301 In *Georgia Today: Facts and Trends.* J. C. Belcher and K. I. Dean, Eds. Athens GA: University of Georgia Press, 1960.

Schrade, B. "Coyotes on Prowl: Blamed in Cobb Animal Deaths." *Atlanta Journal-Constitution*, 29 November 1999.
Schrenk, H. H., H. Heimann, G. D. Clayton, W. M. Gafafer, and H. Wexler. 1949. "Air Pollution in Donora, Pa. Epidemiology of the Unusual Smog Episode of October 1948." Preliminary Report. Public Health Bulletin No. 306. Federal Security Agency, Public Health Service. Washington, DC.
Scott, P. "Growth Flushes Out Hogs in Henry." *Atlanta Journal-Constitution*, 19 June 2000.
Seabrook, C. 1998a. "Whoosh! Peregrines Out of Danger." *Atlanta Journal-Constitution*, 25 August 1998.
———. 1998b. "Local Peregrine Falcon Breaks Free From Endangered Species List." *Atlanta Journal-Constitution*, 26 August 1998.
———. 1999a. "Choking the Life Out of Atlanta." *Atlanta Journal-Constitution*, 20 June 1999.
———. 1999b. "Asthma Is a Threat That's Growing." *Atlanta Journal-Constitution*, 20 June 1999.
———. 1999c. "Tiny Pollutants a Major Health Concern." *Atlanta Journal-Constitution*, 21 June 1999.
———. 2000a. "Earth Day: Progress 30 Years Later?" *Atlanta Journal-Constitution*, 22 April 2000.
———. 2000b. "Seeing Green." *Atlanta Journal-Constitution*, 3 January 2000.
———. 2000c. "Atlanta's Smog Rivals L.A.'s." *Atlanta Journal-Constitution*, 5 May 2000.
———. "Georgia's Disappearing Songbirds: A Three-day Series." *Atlanta Constitution*, 20–22 May 2001.
——— and L. Soto. "Endangered Species Law to Loosen Up." *Atlanta Journal-Constitution*, 19 July 1997.
Seneca, J. J. "Economic Issues in Protecting Public Health and the Environment," 351–75. In *Public Health and the Environment: The United States Experience*. New York NY: The Guilford Press, 1987.
Shaw, S. P. and C. G. Fredine. 1956. "Wetlands of The United States. Their Extent and Their Value to Waterfowl and other Wildlife." US Department of the Interior, Fish and Wildlife Service. Circular 39.
Shearer, L. "Athens' Air: Experts Suspect Ill Wind from Atlanta." *Athens Daily New/Banner-Herald*, 26 September 1999.
Sheffield, R. M. and T. G. Johnson. 1993. "Georgia's Forests, 1989." Southeastern Forest Experiment Station Resource Bulletin S E-133. US Department of Agriculture, Forest Service.
———, N. D. Cost, W. A. Bechtold, and J. P. McClure. 1985. "Pine Growth Reductions in the Southeast." Southeastern Forest Experiment Station Resource Bulletin SE-83. US Department of Agriculture, Forest Service.
Shellenberger, D. L., R. G. Barget, J. A. Lineback, and E. A. Shapiro. 1996. "Nitrate in Georgia's Ground Water." Georgia Department of Natural Resources, Georgia Geological Survey. Project Report 25. Atlanta GA.
Simons, T. R., K. N. Rabenold, D. A. Buehler, J. A. Collazo, and K. E. Franzreb. "The Role of Indicator Species: Neotropical Migratory Song Birds," 187–208. In *Ecosystem Management for Sustainability: Principles and Practices Illustrated by a Regional Biosphere Reserve Cooperative*. Edited by J. D. Piene. Boca Raton FL: CRC Press; Lewis Publishers, 1999.
Sisk, L. *The Changed Look of the Countryside*. Franklin Springs GA: Advocate Press, 1975.
Sloane, C. S. 1984. "Meteorologically Adjusted Air Quality Trends: Visibility." *Atmospheric Environment*. 18:1217–29.
Smith, M. C. and J. Sellers. 1995 "Water Quality Impacts of Poultry Litter Applications," 171–74. In *Proceedings of the 1995 Georgia Water Resources Conference*. 11–12 April 1995. K. J. Hatcher, ed. Athens GA.

Smith, R. A. and R. B. Alexander. 1983. "Evidence for Acid-Precipitation-Induced Trends in Stream Chemistry at Hydrolic Bench-Mark Stations." US Geological Circular 910. US Geological Survey, Washington, DC.

Soto, L. 1999. "Here's the Dirt: Builders May Face Crackdown." *Atlanta Constitution*, 5 April 1999.

Southeast Region Post-War Planning Committee. 1944. "Georgia Post-War Agriculture. Preliminary Report." US Department of Agriculture, Washington, DC.

Spengler, J. and R. Wilson. "Emission, Dispersion and Concentration of Particles," 41–62 In *Particles in Our Air*. R. Wilson and J. Spengler, eds. Cambridge MA: Harvard School of Public Health, 1996.

Spillers, A. R. 1939a. "Forest Resources of Central Georgia." Forest Survey Release No. 41. Southern Forest Experiment Station, US Department of Agriculture, Forest Service. New Orleans, LA.

———. 1939b. "Forest Resources of North Central Georgia." US Department of Agriculture, Forest Service. Forest Survey Release No. 44. Southern Forest Experiment Station, US Department of Agriculture, Forest Service. New Orleans LA.

———. 1939c. "Forest Resources of North Georgia." Forest Survey Release No. 45. Southern Forest Experiment Station, US Department of Agriculture, Forest Service. New Orleans LA.

———. and Eldredge, I. F. 1943. "Georgia Forest Resources and Industries." US Department of Agriculture Miscellaneous Publication 501.

Spraberry, J. A. 1965. "Summary of Reservoir Sediment Deposition Surveys Made in the United States through 1960." US Department of Agriculture Miscellaneous Publication 964. Washington, DC.

St. John, J. C. and W. L.Chameides. 1997. "Climatology of Ozone Exceedences in the Atlanta Metropolitan Area: 1-hour vs 8-hour Standard and the Role of Plume Recirculation Air Pollution Episodes." *Environmental Science Technology*. 31:2791–2804.

State Soil and Water Conservation Commission. 1981. "Georgia Resource Conservation Program and Action Plan." State Soil and Water Conservation Committee.

———. 1994. "Agricultural Best Management Practices for Protecting Water Quality in Georgia." State Soil and Water Conservation Commission, Athens GA.

———. 1995. Georgia's Land: Its Use and Condition. Produced jointly by the State Soil and Water Conservation Commission and US Department of Agriculture, Natural Resources Conservation Service, Athens GA.

Stell, S. M., E. H. Hopkins, G. R. Buell, and D. J. Hippe. 1995. "Use and Occurrence of Pesticides in the Apalachicola-Chattahoochee-Flint River basin, Georgia, Alabama, and Florida, 1960–91." US Geological Survey Open-File Report 95-739.

Stensland, G. J. and R. G. Semonin. 1982. "Another Interpretation of the pH Trend in the United States." *Bulletin of the American Meteorological Society*. 63:1277–84.

Stinner, D. H. "Colonial Wading Birds and Nutrient Cycling in the Okefenokee Swamp." Ph. D. Dissertation, University of Georgia, 1983.

Stoddard, H. L. *The Bobwhite Quail. Its Habits, Preservation, and Increase*. New York NY: Charles Scribner's Sons, 1931.

———. 1939. "The Use of Controlled Fire in Southeastern Game Management." Cooperative Quail Study Association. Thomasville GA.

———. 1963. "Maintenance and Increase of the Eastern Wild turkey on Private Lands of the Coastal Plain of the Deep Southeast." Tall Timbers Research Station Bulletin No. 3. Tall Timbers Research Station, Tallahassee FL.

———. 1975. "Birds of Grady County, Georgia." Tall Timbers Research Station Bulletin No. 21. Tall Timbers Research Station, Tallahassee FL.

Stossel, J. 2001. "Tampering with Nature." ABC News Special Report, 29 June, 2001.

Sweat, D. 1992. "Wild turkey in Georgia: A Report." *Georgia Wildlife*. 2(1):58–59.

Tansey J. B. and N. D. Cost. 1984. "Georgia Biomass Study: Preliminary Results," 37–49. In *Proceedings of the 1984 Southern Forest Biomass Workshop*. US Department of Agriculture Forest Service, Ashville NC.

Tarr, J. A., *The Search for the Ultimate Sink: Urban Pollution in Historical Perspective*. Akron: The University of Akron Press, 1996.

Tate, W. 1960. "Our Cultural Heritage," 51–63. In *Georgia Today: Facts and Trends*. J. C. Belcher and K. I. Dean, eds. Athens GA" University of Georgia Press, 1960.

Tennessee Valley Authority. 1990. "Status of Blue Ridge Reservoir, An Overview of Reservoir Conditions. Tennessee Valley Authority, Water Resources and Ecological Monitoring." Water Resources Division.

Thackston, R., T. Holbrook, W. Abler, J. Bearden, D. Carlock, D. Forster, N. Nicholson, and R. Simpson. 1991. "The Wild Turkey in Georgia: History, Biology, and Management." Georgia Department of Natural Resources. Atlanta GA.

Thompson, M. T. 1998. "Forest Statistics for Georgia, 1997." Resource Bulletin SRS-36. US Department of Agriculture Forest Service, Southern Research Station, Asheville NC.

Todd, R., R. Lowrance, O. Hendrickson, L. Asmussen, R. Leonard, J. Fail, and B. Herrick. 1983. "Riparian Vegetation as Filters of Nutrients Exported from a Coastal Plain Agricultural Watershed." In *Nutrient Cycling in Agricultural Ecosystems*. R. R. Lowrance, R. L. Todd, L. E. Asmussen, and R. A. Leonard, eds. University Georgia Agricultural Experiment Station Special Publication No. 23.

Trijonis, J. 1986. "Patterns and Trends in Data for Atmospheric Sulfates and Visibility," pp. 109–27 In *Acid Deposition-Long-Term Trends*. National Academy Press. Washington, DC.

Trimble, S. W. 1970. "The Alcovy River Swamps: The Result of Culturally Accelerated Sedimentation." *Bulletin of the Georgia Academy of Science*. 28:131–41.

———. *Man-Induced Soil Erosion on the Southern Piedmont 1700–1970*. Alkeny IA: The Soil Conservation Society of America, 1974.

———. 1975. "Denudation Studies: Can We Assume Stream Steady State?" *Science*. 188:1207–1208.

———. 1985. "Perspectives on the History of Soil Erosion in the Eastern United States." *Agricultural History*. 59:162–80.

——— and P. Crosson. 2000. "US Soil Erosion Rates—Myth and Reality." *Science*. 289:248–50.

———, F. H. Weirich, and B. L. Hoag. 1987. "Reforestation and the Reduction of Water Yield on the Southern Piedmont since Circa 1940," *Water Resources Research*. 23:425–37.

Troxler, R., D. Reinhart, and A. Hallum. 1983. "Metro Atlanta Water Pollution Control—A Decade of Progress." *Journal of the Water Pollution Control Federation*. 55:1121–27.

Turner, M. G. 1987. "Land Use Changes and Net Primary Production in the Georgia, USA, landscape: 1935–1982." *Environmental Management*. 11:237–47.

——— and C. L. Ruscher. 1988. "Changes in Landscape Patterns in Georgia, USA." *Landscape Ecolology*, 1:241–51.

Tyson, A. W., P. Bush, R. Perkins, and W. Segars. 1995. "Nitrate Contamination of Domestic Wells in Georgia," 142–45a. In *Proceedings of the 1995 Georgia Water Resources Conference*. 11–12 April 1995. K. J. Hatcher, ed. Athens GA.

US Army Corps of Engineers. 1998. "Analysis of Reservoir Storage Loss at West Point Lake, Chattahoochee River, Georgia and Alabama." Prepared for the US Army Corps of Engineers, South Atlantic Division, Mobile District by HDR Engineering Inc., Atlanta GA.

———. 2000. "Sedimentation Analysis of Allatoona Lake, Etowah River, Georgia." Prepared for the US Army Corps of Engineers, South Atlantic, Mobile District by HDR Engineering Inc., Marietta GA.

US Bureau of the Census. 1920–2000. *Census of Agriculture*. US Government Printing Office, Washington, DC.

US Department of Agriculture. 1946. "Basic Data on Forest Area and Timber Volumes from the Southern Forest Survey, 1932–1936." Forest Survey Release No. 54. Southern Forest Experiment Station, US Department of Agriculture Forest Service. New Orleans LA.

———. 1950. "Report of the Chief of the Soil Conservation Service." In Annual Report of the US Department of Agriculture. US Printing Office, Washington, DC.

———. 1960–1970. Annual Reports on Soil and Water Conservation in Georgia. US Department of Agriculture, Soil Conservation Service. Athens GA.

———. 1961a. "Technical Memorandum on Drainage and Reclamation to the United States Study Commission, Southeast River Basins." Basins 1-7. Prepared by the US Department of Agriculture, Soil Conservation Service. Washington, DC.

———. 1961b. "Technical Memorandum on Sediment Control to the United States Study Commission. Southeast River Basins, Alabama—Florida —Georgia —North Carolina— South Carolina. Basin No. 3." Prepared by the US Department of Agriculture, Soil Conservation Service. Washington, DC.

———. 1977. "Cropland Erosion." Prepared by the US Department of Agriculture, Soil Conservation Service for use in Water Resources Council's Second National Water Assessment. Washington, DC.

———. 1993. Final Report of the Georgia Watershed Agricultural Nonpoint Source Pollution Assessment. Cooperative River Basin Study. Prepared by US Department of Agriculture Forest Service and Soil Conservation Service. Sponsored by Georgia Soil and Water Conservation Commission, and the Georgia Department of Natural Resources, Environmental Protection Division.

———. 1994. "1992 National Resources Inventory." National Resources Conservation Service, Washington, DC.

———. 1997. "Agricultural Resources and Environmental Indicators, 1996–97." Agricultural Handbook No. 712. US Department of Agriculture, Economic Research Service, Natural Resources and Environmental Division. Washington, DC.

———. 1999. "1997 National Resources Inventory." US Deptartment of Agriculture, National Resources Conservation Service, Washington, DC.

US Department of Commerce. 1973–1980. "Georgia Landings, Annual Summary." National Oceanic and Atmospheric Administration, National Marine Fisheries Service. Washington, DC.

US Department of Health, Education and Welfare. 1958. "Air Pollution Measurements of the National Air Sampling Network. Analyses of Suspended Particulates, 1953–1957." US Public Health Service. US Government Printing Office, Washington, DC.

———. 1962. "Air Pollution Measurements of the National Air Sampling Network. Analyses of Suspended Particulates, 1957–1961." US Public Health Service. US Government Printing Office, Washington, DC.

US Department of the Interior. 1966. "Proceedings of the Conference: In the Matter of Pollution of the Interstate Waters of the Chattahoochee River and Its Tributaries, from Atlanta, Georgia to Fort Gaines, Georgia." Atlanta GA. 14–15 July 1966. 2 volumes. US Department of the Interior, Federal Water Pollution Control Administration, Washington, DC.

———. 1969. "Proceedings of the Conference: In the Matter of Pollution of the Interstate Waters of the Lower Savannah River and Its Estuaries, Tributaries and Connecting Waters- Georgia-South Carolina." Second Session, Savannah GA. 29 October 1969. US Department of the Interior, Federal Water Pollution Control Administration, Washington, DC.

———. 1970. "Proceedings of the Conference: In the Matter of Pollution of the Interstate Waters of the Chattahoochee River and Its Tributaries—Georgia—Alabama." Atlanta GA. 17 February 1970. US Department of the Interior, Federal Water Pollution Control Administration, Washington, DC.

———. 1989. "1985 National Survey of Fishing, Hunting, and Wildlife-Associated Recreation: Georgia." US Department of the Interior, Fish and Wildlife Service. Washington, DC.

———. 2000a. "Meetings Will Discuss Resident Canada Goose Conflicts." US Fish and Wildlife Service News Release, 4 January 2000.
———. 2000b. "President Clinton Signs Bill Authorizing Seed Money for Neotropical Birds." US Fish and Wildlife Service News Release, 3 August 2000.
———. 2000c. "President Clinton Urges Americans to Celebrate Eighth Annual International Migratory Bird Day." US Fish and Wildlife Service News Release, 12 May 2000.
——— and US Department of Commerce. 1998. "1996 National Survey of Fishing, Hunting, and Wildlife-Associated Recreation, Georgia." US Department of the Interior, Fish and Wildlife Service, and US Department of Commerce, Bureau of the Census. Washington, DC.
US Department of Transportation. 1981. Volume 11 of Wind-Ceiling-Visibility Data At Selected Airports. "Visibility Time Trends for Key Stations." Federal Aviation Administration, Plans Development Division, Office of Aviation Policy and Plans.
US Environmental Protection Agency. 1973–1977. "National Emissions Reports." Office of Air and Waste Management. Office of Air Quality Planning and Standards, Research Triangle Park NC.
———. 1973–1998. "National Air Quality and Emissions Trends Report." Office of Air Quality Planning and Standards. Research Triangle Park NC.
———. 1980. "National Accomplishments in Pollution Control: 1970–1980." US Environmental Protection Agency, Washington, DC.
———. 1992. "A Citizens Guide to Radon (Third edition): The Guide to Protecting Yourself and Your Family from Radon." Office of Air and Radiation, Indoor Environments Division. EPA Document 402-K92-001. From the EPA website, http://www.epa.gov/iaq/radon/pubs/citguide.html
———. 1994. "National Air Pollutant Emission Trends, 1900–1994." EPA Office of Air Quality Planning and Standards, Research Triangle Park NC.
———. 1994–2001. "Aerometric Information Retrieval System (AIRS)." Air pollution data online at the US EPA website http://www.epa.gov/airs/
———. 1997. "Mercury Study Report to Congress." Volume 1. Executive Summary. SAB Review Draft. EPA Office of Air Quality Planning and Standards and Office of Research and Development.
———. 1996. "Air Quality Criteria for Particulate Matter." Volume 1. National Center for Environmental Assessment. Office of Research and Development. Research Triangle Park NC.
———. 1999a. "Summary of the 2000 Budget." Office of Chief Financial Officer. January 1999. Washington DC.
———. 1999b. "The National Survey of Mercury Concentrations in Fish. Data Base Summary, 1990–1995." US EPA, Standards and Applied Science Division. Washington, DC
———. 1999c. "1997 Georgia Toxics Release Inventory Report." USEPA Envirofacts Warehouse webpage www.epa.gov/enviro/html/tris/tris_state.html
US Food and Drug Administration. 2001. "Mercury In Fish: Cause For Concern?" FDA Consumer, September 1994; Revised May 1995. From the FDA webpage http://www.fda.gov.
US Geological Survey. 1999. "The Quality of Our Nation's Waters–Nutrients and Pesticides." US Geological Survey Circular 1225. Washington, DC.
Van Kessel, J. F. 1977. "Removal of Nitrate from Effluent Following Discharge on Surface Water." *Water Research* 11:533–37.
Wade, D. D. and J. D. Lunsford. 1989. "A Guide for Prescribed Fire in Southern Forests." US Department of Agriculture, Forest Service, Southern Region. Technical Publication R8-TP 11.
Walker, J. T. 1982. "Characterization of Rain in the Georgia Piedmont and Effects of Acidified Water on Crop and Ornamental Plants." Agricultural Experiment Station Research Bulletin 283. University of Georgia, Athens GA.
Walker, J. T. and J. B. Melin. 1998. "Thirteen Years of Monitoring Rain and Atmospheric Deposition at Three Georgia Locations 1984–1996." Georgia Agricultural Experiment Station. Research Report No. 652. University of Georgia, Athens GA.

Werblow, D. A., and F. W. Cubbage. 1985. "An Analysis of Tree Planting Investments in Georgia under the Proposed Conservation and Reforestation Act." Working Paper No. 9. Southeastern Center for Forest Economics Research. Research Triangle Park, North Carolina.

West, D. C., T. W. Doyle, M. L. Tharp, J. J. Beauchamp, W. J. Platt, and D. J. Downing. 1993. "Recent Growth Increases in Old-growth Longleaf Pine." *Canadian Journal of Forest Research*. 23:846–53.

Wetherington, M. V. *The New South Comes to Wiregrass Georgia; 1860–1910.* Knoxville TN: University of Tennessee Press, 1994.

Weyandt, T. L. Jr. 1998. "Vehicle Registrations in the Metropolitan Atlanta Region." Research Atlanta, Inc. and The School of Policy Studies, Georgia State University, Atlanta GA.

Wharton, C. H. *The Southern River Swamp–A Multiple-Use Environment.* Atlanta: Georgia State University, 1970.

Wharton, C. H. *The Natural Environments of Georgia.* Atlanta: Georgia Department of Natural Resources, 1997.

White, D. H., C. B. Kepler, J. S. Hatfield, P. W. Sykes, Jr., and J. T. Seginak. 1996. "Habitat Associations of Birds in the Georgia Piedmont during Winter." *Journal of Field Ornithology*. 67:159–66.

Whitney, G. G. *From Coastal Wilderness to Fruited Plain; A History of Environmental Change in Temperate North America from 1500 to Present.* Cambridge UK: Cambridge University Press, 1994.

Williams, C. 1999. "Oh, Deer: Animal-auto Collisions Add Up." *Atlanta Constitution*, 15 November 1999.

Wilson, R. Introduction. pp. 1–14 In *Particles in Our Air.* R. Wilson and J. Spengler, eds. Cambridge MA: Harvard School of Public Health, 1996.

Winn, B. 1968. "The Fallow Fields." *Atlanta Magazine.* 7(11):85–89, 104, 106.

Works Progress Administration. *Georgia: The WPA Guide to Its Towns and Countryside.* Compiled by Workers of the Writer's Program of the Works Progress Administration in the State of Georgia. Columbia: University of South Carolina Press, 1990.

Wright, A. H. and F. Harper. 1913. A Biological Reconnaissance of Okefenokee Swamp: The Birds. *Auk.* 30:477–505.

Xiao-Qing, Z. and T. C. Rasmussen. 1999. "Water Quality Dynamics in Lake Lanier." pp. 350–53 In *Proceedings of the 1999 Georgia Water Resources Conference.* 30–31 March 1999. K. J. Hatcher, ed. Athens GA.

Young, I., J. Gholson, and C. N. Hargrove. *History of Macon, Georgia.* Macon GA: Lyon, Marshall and Brooks, 1950.

Zahner, R., J. R. Saucier, and R. K. Myers. 1989. "Tree-ring Model Interprets Growth Decline in Natural Stands of Loblolly Pine in the Southeastern United States." *Canadian Journal of Forest Research.* 19:612–21.

Zeleny, L. *The Bluebird; How You Can Help Its Fight for Survival.* Bloomington IN: Indiana University Press, 1976.

Zeller, H. D. and H. N. Wyatt. Selective Shad Removal in Southern Reservoirs. pp. 405–14. In *Reservoir Fishery Resources. A Symposium of the Reservoir Committee of the Southern Division, American Fisheries Society.* Athens GA: Southern Division, American Fisheries Society, 1967.

INDEX

acid rain 214-223
Aderhold, O. C. 92
air pollution 189-252, 12, 335
Alamogordo, NM 160
Albany, GA 174, 206, 212, 265
Albany By-pass Pond 172
Albany Journal 273
Alcovy River 44, 71, 99
Allied Chemical Corporation 161
alligator 283, 323
Altamaha River 63, 308
Ambient Air Surveillance Report 199
American Forestry Association 39
anions 216-219
aquifer 148
armadillo 325
Arnold, Ellis 79
Associated Press 309, 320
asthma 191, 250
Athens, GA 3, 78
Athens Daily News-Banner Herald 230
Atlanta, GA 5, 54, 108, 153, 204
Atlanta Airport 245-249, 289
Atlanta Constitution 5, 54, 82, 94, 125, 177, 230
Atlanta Journal 194, 198, 230
Audubon, John James 290
Augusta, GA 207, 228, 248, 302

Babbitt, Bruce 172
bald eagle 172, 287, 299, 321, 327
Barnes, Roy 55
Barnett Shoals 74
Barrow County 82
Bartow County 102
Bartram, William 62
bass, black 312
bass, largemouth 163, 165
bear 283
beaver 49, 278
Bellville, GA 216
Bennett, Hugh Hammond 63, 82, 85, 92, 116
Big Haynes Creek 41, 43
Big Sandy Creek 256
biochemical oxygen demand (BOD) 133-138
bird sanctuaries 290
birds, song 290-303
Blue Ridge Lake 100, 178, 311
Blue Ridge Mountains 223, 245
bluebird, eastern 294
bobcat 166, 280
Bobwhite Quail Initiative 273
boll weevil 21
Bolton Mill Creek 100
Breeding Bird Survey 271, 292
briar patch 257
Broad River 18, 65, 100
Brooks, Mrs. Richard P. 90
brown-headed cowbird 328
Brunswick, GA 161, 205
Buckhead 264
Buckhorn Creek 119, 256

buffalo 16, 17, 62, 257
Bulloch County 152
Bureau of the Census 43, 47, 232
Burleigh, Thomas 78

calcium 219
Callaway, Cason J. 81
calomel 158
carbon dioxide 243
carbon monoxide 239-242
carboxyhemoglobin 240
Carson, Rachel 6, 171, 335
Carter, Jimmy 9, 53, 89, 112, 198, 266
catfish 311
cations 216-219
Chappell's Millpond 256
Chatham County 199, 213, 226
Chattahoochee National Forest 260, 324
Chattahoochee River 5, 73, 77, 106, 110, 125, 154, 307
Cherokee County 102
chlor-alkali 159
chlordane 171, 173
Christmas Bird Count 271, 286, 297, 327
chromium 168
Civilian Conservation Corps 35
Clarke County 73
Clarks Hill Lake 97, 102, 164, 314
Clayton County 248
Clean Air Act 12, 201, 205
Clinton, Bill 6, 115, 298
Clyo, GA 97

coal 193
Coastal Plain 19, 29
Cobb County 283
Columbus, GA 228, 238
Conasauga River 154
Confederate Avenue 225
Conservation Reserve
  Program 35
conservation tillage 88
Cooperative Quail
  Investigation 273
Coosa River 81, 103, 168
Cope, Channing 82
Copper Hill, TN 220
Cornelia, GA 222
Cornish Creek 71
cotton 21, 22, 27, 62
cotton gins 31
Council on Environmental
  Quality 11, 115
coyote 289, 328
blue crab 162
criteria pollutants 201, 251
crop productivity 114
crops, row 26-28
crow 293
Cumming, GA 283
Cyclorama 91

Dahlonega, GA 159
Dawsonville, GA 216
DDT 9, 52, 171, 295, 327
Decatur, GA 231
deer 16, 259-265
Dekalb College (Georgia
  Perimeter College) 192,
  227
Dekalb County 226, 248
Denmark, Dr. Leila 189,
  193
dieldrin 176
diversification 26, 31
Dixie Crusaders 39
Dodge County 24, 51, 111
Donora, Pennsylvania 190,
  203
Dooly County 153
Dougherty County 212
dove, mourning 268, 273-
  275
drainage 47
Dublin, GA 121

ducks 286, 299
dust 197
dysentery 2, 121

Eastman, GA 123
Eatonton, GA 216
egret, cattle 300
Ellenwood, GA 264
Emanuel County 4
endangered species 283,
  304, 322
Endangered Species Act 322
erosion 59-70, 110, 112
Etowah River 97, 103, 155
eutrophication 177

Fairburn, GA 125
Fall Line 86
Falling Creek 221
Fannin County 226
farms 26-30
farmstead 53
fecal coliform bacteria 129-
  133, 183
fence rows 257
ferries 75
fertilizer 145, 310
fish 306-315
fish advisories 165, 166
fish kills 307
Fish Tissue Assessment
  Project 169
Flat Creek 71, 182
Flint River 15, 75
Ford, Henry 260
forest fire 38-41, 303
forest productivity 36-38
forest regeneration 34, 111
forests 31-41
Fort Valley, GA 150
fossil fuels 223
fox 166, 282
Franklin County 43, 48
Franklin, Benjamin 265
Fuel efficiency 232
Fulton County 108, 201,
  208, 226, 240, 248

geese, Canada 284
Georgia BASS Chapter
  Federation, Inc 313

Georgia Conservancy 199,
  213
Georgia Department of
  Natural Resources 104,
  124, 181, 201
Georgia Environmental
  Protection Division
  (EPD) 10
Georgia Forestry
  Commission 32, 36, 41
Georgia Geological Survey
  42, 148
Georgia Power Company
  104, 137
Georgia Tech 226
Georgia Water Quality
  Control Board 124, 161
Gilmer, George Rockingham
  65
global warming 6
Glynn County 3, 216, 226
good news 1, 9-13
Gore, Al 6
Grady, Henry 116
Great Depression 22
Greene County 22, 81
greenhouse effect 224
Griffin, GA 216
groundwater 146-153
gull 302
gullies 76
Gum Branch Creek 144
Gwinnett County 43, 73,
  83, 226

Hall County 94
Harris County 64
Hawkinsville, GA 121
health 249-251
Henry County 248
Herty, Charles 39
Hiawassee, GA 216
hookworms 2
Hunters for the Hungry 265
hunting 258, 261

Ichawaynochaway Creek 93
Indians 8, 16-18
industrial pollutants 157-
  171
inversions 224

Index

Jackson County 43, 73
Jackson, Henry Rootes 66, 333
Jasper County 81
Jefferson, Thomas 64
Jones County 65, 81

Kellogg Task Force 140, 145
Kennesaw Mountain National Battlefield Park 289
Kennesaw State University 177
Kundell, James 46, 91, 205, 209, 216, 231, 245, 291

Lagrange, GA 106
Lake Acworth 132, 285
Lake Allatoona 98, 102, 156, 177
Lake Blackshear 168
Lake Burton 74
Lake Harding 168, 178
Lake Hartwell 102, 127
Lake Jackson 68, 100, 103, 178
Lake Lanier 128, 168, 179, 182, 285
Lake Nottely 100
Lake Rabun 74
Lake Russell 102
Lake Seminole 164, 284
Lake Walter F. George 308
lakes 127
landfills 197
landscape 30
Lanier, Sidney 65, 77, 115, 186, 333
Laurens County 17, 19
lead 168, 236-239
lead in gasoline 236, 238
lead poisoning 236-239
Leaf, GA 222
Lee County 15
Leopold, Aldo 15, 56
life span 2
Lincoln County 93
Little River 150
livestock 27
London, UK 190, 203
Lumpkin County 289
lung cancer 250

Lyell, Sir Charles 63

Macon, GA 124, 249, 274, 286, 302
Madison County 25, 34, 73
magnesium 219
malaria 2, 121
manure 151
Marietta, GA 289
Mauldin's Mill 72
McCaysville, GA 212, 220
McDuffie County 216
McIntosh, William 17
mercurochrome 158
mercury 158-167, 336
merthiolate 158
methemoglobinemia ("blue baby" syndrome) 139
methylmercury 159
metro Atlanta 108, 153, 226, 231, 264
Milledgeville, GA 212
Millen, GA 288
millponds 73
Montgomery County 19
Morgan Falls Dam 74
Muckafoone Creek 75
Muckafoone Lake 74
Mud Creek 146
Mulberry River 72, 94
Myrick, Susan 82

Nancy Creek 110
neotropical migrant birds 298
news media 5
Nickajack Lake 303
nitrate 138-153, 185
nitrogen 138-153
nitrogen oxides 223, 234
nongame species 287-290
nonpoint source pollution 88, 146

Ochlockonee River 165, 309
Ocmulgee River 16, 18, 125, 283
Oconee River 3, 16, 18, 71, 125
Odum, Eugene 114, 263, 291, 304

Ogeechee River 18, 143, 165, 260
Oglethorpe County 4, 94
Okefenokee National Wildlife Refuge 297, 324
Oglethorpe, General James 18
Okefenokee Swamp 44, 181, 216
Olin Corporation 161, 162
Olmsted, Frederick Law 64
opossum 166, 256
osprey 289
otter 166
oxygen, dissolved 133-138
oyster 317
ozone 223-234
ozone alert 230

particles in air 195, 202-209
pasture 30
Peachtree Creek 110, 132, 143
peregrine falcon 172, 287, 299, 327
pesticides 171-176
pH of rain 214-222
phosphorus 5, 139, 153-157, 185
Piedmont 63, 67, 91
Piedmont National Wildlife Refuge 286
plume recirculation 227
PM-10 195, 203
PM-2.5 195, 203
PCBs 157, 168
ponds 104, 127
population 50-56
primary standards 201
Providence Canyon 59, 68, 86, 94
Pulaski County 19
pulpwood 36
Purvis Creek 161
Putnam County 81

quail (northern bob-white) 268-273, 287

rabbit 276
raccoon 166, 282
reservoirs 183

Resources Conservation Act 114
Richmond County 149
Rivers, E. D. 79
roads 88
Rockdale County 226
Rocky Creek 256
Rome, GA 212
Rooty Creek 150
Rossville, GA 204
Roswell, GA 231
rotenone 312
rough fish 311
rural population 50-52

Sandersville, GA 205
Sandy Creek 25, 73, 74, 80, 99
Sandy Springs, GA 283
Satilla River 136, 165
Savannah, GA 18, 205, 248
Savannah River 97, 125, 127, 163, 307
screw worm 335
Scull Shoals 71, 92, 96
Seabrook, Charles 172, 229, 291, 299
seafood 315-321
sediment 89, 106, 125-128, 167
sedimentation 93, 99-106, 112, 183
settlers 9
sewage 128-133, 319
shad, American 306, 317
shad, gizzard 312
shad, threadfin 312
sharecropper 23, 24, 115
shellfish 317, 319
shrimp 316
Silent Spring 6, 171
small game 277
smog 224, 229
Smokey Bear 39, 304
Snake Creek 110
Soap Creek 93
Soil Bank 32, 34, 114
soil conservation 77-92
Soil Conservation Districts 79, 87
Soil Conservation Service (SCS) 47, 71, 73

Soil Erosion Service 80
songbirds 290-303
Sope Creek 174, 308
Soque River 110
South River 71, 132, 143, 155
sparrow, Bachman's 299
sparrow, field 294
sparrow, house (English sparrow) 294, 296
squirrel 276
starling 296
Stephens County 48
Stewart County 68
Stilesboro, GA 220
stork, wood 288, 323
sulfur dioxide 209-213
Summerville, GA 216
swamps 41
swordfish 159

Tallulah River 222
Talmadge, Eugene 79
Talmadge, Herman 79, 82
technology 335
Telfair County 19
tenants 23, 24
Thomas County 37, 301
Thomas, C. C. 3
Thomasville, GA 269, 310
Thompson, M. E. 82
thrasher, brown 257, 290
Thurmond Lake 97, 102
tillage 87
Tilton, GA 216, 283
tobacco smoke 240, 250
trees 233
Trimble 18, 61, 71, 78, 85, 91, 111, 125
Trophic State Index 178
tuna 159
turkey, wild 265-267, 299
Turner Creek 69
Turtle River 161
typhoid 2, 121

U. S. Army Corps of Engineers 102
U. S. Department of Transportation 246

U. S. Environmental Protection Agency (EPA) 10
U. S. Forest Service 32, 108
U. S. Geological Survey 103, 124, 147, 154, 167
U. S. Public Health Service 122, 192
ultra-violet light 194
University of Georgia 24, 77, 216
urban growth 49-56, 110
urban sprawl 232

visibility 245-249
vitamin D 194
volatile organic compounds (VOC) 223, 233

Walton County 43
wastewater 138
water quality index 176
water wells 146-153
Waycross, GA 216
West Point Lake 106, 107, 168, 178, 183, 310, 314
wetlands acreage 41-49
Wharton 44, 49, 71
Whitfield County 81
Winder, GA 82
wiregrass region 20
wood (as a fuel) 195
woodpecker, red-cockaded 304
Woody, Arthur 260
Works Progress Administration (WPA) 68
World War II 8, 26, 49, 115, 121
wren, Bewick's 302
wren, house 296
Wrens, GA 207

Yellow River 71, 72

zinc 168